T0280187

THE GREAT ARCHITECTS
OF MARS

"The pyramids, buildings, and megalithic reliefs that George Haas has uncovered in this book leave little doubt that an advanced race of beings once inhabited Mars. In the face of this evidence, it's curious why NASA finds it necessary to ignore and cover up this vital information and the unknown history that waits to be explored."

CLIFF DUNNING, AUTHOR AND
HOST OF *EARTH ANCIENTS PODCAST*

"George Haas has written a thorough, well-researched new tome on the Mars and Cydonia question that adds new detail, makes new discoveries, and enhances the debate on these long-disputed subjects. Highly recommended and a worthy addition to a discourse that now, more than ever, needs to be engaged regarding the Red Planet."

MIKE BARA, AUTHOR OF *ANCIENT ALIENS ON MARS*

"It's high time the hypothesis of an ancient civilization on Mars be given serious attention by the scientific community. George Haas is to be commended for undertaking his analysis and presenting his findings of potential Martian structures."

M. J. CRAIG, AUTHOR OF *SECRET MARS*

"George Haas analyzes unusual and seemingly artificial features on the surface of Mars from an artistic and architectural perspective. He compares these features with similar ones on Earth and builds a case for the development of an ancient civilization on Mars not unlike our own. These sites may be good targets of opportunity for future exploration by planetary probes and human expeditions."

MARK CARLOTTO, AEROSPACE ENGINEER AND
AUTHOR OF *BEFORE ATLANTIS*

"Very interesting indeed, particularly the comparisons of the North American Native mound complexes with the pyramidal formations on Mars. The pictures are clear; the arguments are not overzealous. The book will somberly and persuasively add to the dialogue and debate about life on Mars—past and present."

ANANDA SIRISENA, PRESIDENT OF
THE SOCIETY FOR PLANETARY SETI RESEARCH

"There are a lot of geoglyphs on Mars that seem to resemble things on Earth, and many of them have some similarity to different things that were in Mesoamerica and various Mesopotamian cultures. It's worth exploring."

FRANK MORANO, HOST OF
THE OTHER SIDE OF MIDNIGHT

THE GREAT ARCHITECTS
OF MARS

Evidence for the Lost Civilizations on the Red Planet

A Sacred Planet Book

GEORGE J. HAAS

Bear & Company
Rochester, Vermont

Bear & Company
One Park Street
Rochester, Vermont 05767
www.BearandCompanyBooks.com

Bear & Company is a division of Inner Traditions International

Sacred Planet Books are curated by Richard Grossinger, Inner Traditions editorial board member and cofounder and former publisher of North Atlantic Books. The Sacred Planet collection, published under the umbrella of the Inner Traditions family of imprints, includes works on the themes of consciousness, cosmology, alternative medicine, dreams, climate, permaculture, alchemy, shamanic studies, oracles, astrology, crystals, hyperobjects, locutions, and subtle bodies.

Cataloging-in-Publication Data for this title is available from the Library of Congress

ISBN 978-1-59143-516-7 (print)
ISBN 978-1-59143-517-4 (ebook)

Printed and bound in India by Nutech Print Services

10 9 8 7 6 5 4 3 2 1

Text design and layout by Debbie Glogover
This book was typeset in Garamond Premier Pro with Atrament, Gill Sans MT Pro, and Real Head Pro

To send correspondence to the author of this book, mail a first-class letter to the author c/o Inner Traditions • Bear & Company, One Park Street, Rochester, VT 05767, and we will forward the communication, or contact the author directly at **https://thecydoniainstitute.com**.

Scan the QR code and save 25% at InnerTraditions.com.
Browse over 2,000 titles on spirituality, the occult, ancient mysteries, new science, holistic health, and natural medicine.

Destroy not the ancient landmarks that thy fathers set up.

PROVERBS 22:28

Contents

The Second Copernican Revolution

John E. Brandenburg, Ph.D.

THE PLANET MARS has been a driver of scientific progress since the foundations of humanity. Now we are investigating the planet with powerful tools, and it has become the focus of our future human space exploration. We are deep in the scientific endgame at Mars regarding the great questions that have haunted humanity since it invented the scientific process: Are we a part of a community of peoples in the cosmos, and are living planets like Earth common in the universe? Mars, the first Earthlike planet we can visit in the vast cosmos, appears to hold the answer to these questions.

This book, documenting extraordinary research done by George Haas over many years, is showing us that emerging answer. The answer appears to be yes and yes: yes, we are part of a community of peoples in a living cosmos, and yes, living planets like Earth are common in the universe. Unfortunately, it was Mars that emerged as the second living planet in the solar system, before it suffered some nameless calamity. Mars, as will be shown in the pages that follow, was once the home of people like us, who like our ancestors built great edifices to awe those who would see them in the future. The planet Mars and what we are finding there, due to great efforts of the author of this book and others, are forming the basis for a second Copernican Revolution: the discovery that we are not only not the center of the universe geometrically, but we are not the center of the intelligent biological universe as well.

Mars, with its red color and its unique path, spanning the heavens with its dramatic variance in brightness, has fascinated ancient stargazers who associated its attributes with war. This made Mars the focus of attention for well-paid court astrologers across the ancient world who counseled kings on its movements, both present and future, so those kings and monarchs would know when to engage their neighbors in tests of military power. Great Empires were formed and failed based on the interpretations of Mars's movements. So a great portion of human history was inspired by Mars's influences.

Mars became the inspiration for the scientific revolution. Its orbit was distinctly elliptical and hence did not fit the basic Copernican Model, which while basically correct, gave all the planets perfectly circular orbits. The elliptical orbit of Mars was not only noted by the ancient Maya but by Tycho Brahe, a court astrologer to a minor prince in central Europe. It was Tycho who commended its study to his deputy Johannes Kepler. From careful study of the movements of Mars, Kepler deduced the laws of planetary motion, which later inspired Isaac Newton to discover gravitation, the laws of motion, and calculus. Another of Newton's advances, a deeper understanding of optics, enabled later Victorian scientists to examine Mars with ever more powerful telescopes.

The telescopic exploration of Mars led immediately to the discovery of its Earthlike environment. Early observations revealed a clear but stormy atmosphere, solid surface, and polar caps, making it more like our home world than any other planet in our solar system. And as telescopes improved, both Giovanni Schiaparelli and Percival Lowell observed what they thought was evidence of intelligent life: linear features they called canals. Because of these developments, a consensus developed among the educated that Mars was Earthlike in detail, being the home of not just life but intelligence. This conviction and its effect upon human culture resulted in the invention of the liquid-fueled rocket by Konstantin Tsiolkovsky and Robert H. Goddard so we could send probes and ultimately people to Mars to find out if indeed Mars held a civilization, past or present.

Now, because of the scientific advances inspired down through the ages by a beckoning Mars, we are in the endgame of our great quest. Because of the rocket technology developed by Wernher von Braun and Goddard so we could get to Mars, and because of the optical technology created by Newton, we can observe Mars's surface from orbit in great detail and send land probes to explore its soil and landscapes.

It has now been found that NASA, in the 1963 Rand study that laid the groundwork for its founding, expected that relics of extraterrestrial civilizations would be discovered in the process of exploring the solar system. News of such discoveries was anticipated by the Brookings Institution in a 1960 report to have a devastating effect on human society, and such news was therefore to be suppressed. Therefore, NASA fully expected that close observation of Mars might find evidence of archeology on its surface, and in fact that is what was found.

Beginning with the Mariner 9 probe that orbited Mars in 1971, it discovered the Pyramids of Elysium and imaged them twice at great difficulty and at high resolution. This was followed by the dramatic discovery of the Face of Mars by the Viking 1 orbiter, at the Prime Viking Landing site in Cydonia, a discovery that NASA confirmed, in secret, by bringing the spacecraft back over the site thirty-five orbits later to take a second series of images. Now with the Face at Cydonia and the pyramidal formations at Elysium confirmed and massive evidence for Earthlike conditions, even life on an early Mars, we must now look at everything with new eyes.

George Haas has a history of presenting great discoveries on Mars to the public, such as the Parrot Geoglyph in the Argyre Basin and the keyhole-shaped landform in the Libya Montes region that bears a startling resemblance to the massive "keyhole" burial mound located in Kofun, Japan. He has investigated an array of structural formations that display a high level of geometric symmetry across the entire planet. If he is correct, Mars was once the home of vibrant and artistic culture and an Earthlike diverse biosphere, predating any such development on Earth. This is a unique and extremely important book that provides the clearest and most precise evidence ever compiled to support the existence of artificial structures on the surface of Mars. These discoveries, if confirmed, will launch the Second Copernican Revolution, and we will never look at Mars, or the night sky, the same way again. So read on, curious reader, and see what this new interpretation of Martian landforms will tell us.

JOHN E. BRANDENBURG, PH.D.

JOHN E. BRANDENBURG, PH.D., is a theoretical plasma physicist and astrophysicist. He received his Ph.D. from the University of California at Davis California in 1981. He is a member of the Society for Planetary SETI Research.

The Question of Objective Existence

James S. Miller

WHEN IT COMES TO MARS, the public and scientists have many questions that bring to mind the painting by Paul Gauguin titled: *Where Do We Come From? What Are We? Where Are We Going?* Like Gauguin, we also want to know the "Where" and "What," but most importantly we want answers to know "Who," "Why," and "How."

These are the questions always asked whenever any kind of investigation is undertaken. It may be an overwhelmed mother trying to find "Who" broke the vase in the dining room. It may be a reporter or a detective coming to the scene of a crime trying to figure out "How" to unravel the sequence of events. Or it may be an archaeologist uncovering an ancient site and trying to find out "Where" this unknown civilization came from.

As we research the planet Mars, whether it be the long-range images from orbiting cameras or the up-close examinations provided by the various rovers over the years, we use the same process and have the same questions. The reality is we cannot answer most of them.

We are pigeonholed into surmising most of what we see, simply because we cannot get firsthand access to the objects themselves. Once on-site we will confirm many of the objects, I am certain. But until we land on Mars and achieve ground truth, we must accept the limitations of distance and instrumentation.

At this point we can easily answer the "Where;" it's Mars obviously. We can speculate on the "How:" these formations are either constructed from scratch or are the result of manipulation of existing terrains. Both are done as a normal process here on Earth. It shouldn't be much different on Mars or elsewhere. The "When" has certain parameters based on how eroded an area appears. The "What" is the first speculation we must engage with. We know we are looking at figurative images of humanoid or animal formations and even some with geometric and pyramidal shapes. But we cannot state it is 100 percent true; it is only what it looks like in the images.

Now to the "Who" and "Why" of the discussion . . . Well, there are no real answers, only speculations. Here on Earth, we have recently discovered thousands of geoglyphs in the exposed jungles of the Amazon and massive ruins within a hill at Göbekli Tepe, and we have no answers as to "Who" built them or "Why." Archaeologists tend to end up with ceremonial or religious reasons when all else eludes them. None have even come up with a valid theory as to the "Who" and "Why" for these Earthbound places. Nor have they established "How" any of it was actually constructed. Even though Göbekli Tepe has been found to be twelve thousand years old, archaeologists still cling to the five-thousand-year-old theory of the origins of modern civilization. So can we even depend on them when looking at Mars?

With so many sites on Earth left unexplained, any speculation on the origins of these formations on Mars just gets that much harder. The only thing we know for sure is that these objects exist and need explanation. Similar questions caught the attention of SETI member and space archaeologist Kathryn Denning. When considering the possibility of finding material remains on another planet, she proposes the questions "What do Others know of their worlds? What do They do there? How can We learn about Them?—are the same. It is not surprising, therefore, that anthropology, archaeology, and SETI share certain core issues."[1]

After many years of investigation and uncovering similarities in the style and design of these objects on Mars, the author of this book believes the time for speculation is over. Like Gauguin we are on a quest to find a lost paradise that will confirm or challenge our fragile beliefs. That lost paradise might just be found on Mars.

JAMES S. MILLER

JAMES S. MILLER is an image analyst and the founder of the Mars research group The Anomaly Hunters.

Preface

Moving us to Mars, past the satellites and stars

COLDPLAY, "MOVING TO MARS"

BEFORE I GET STARTED, I would like to be clear about my background and qualifications. I am not a scientist. I'm an artist. My early education was in the Visual Arts, which included painting, sculpture, and photography. I'm more of a right-brain, free-thinking, creative person as opposed to the left-brain, hard-edged academic. As an artist I have studied the visual wonders of the world around me, and although I may not "know much about trigonometry," I do know artwork when I see it. And Mars—it is filled with artwork.

During the 1980s I was an art instructor and exhibited extensively throughout the New Jersey and New York area and eventually became the director of the Sculptors' Association of New Jersey. I made monthly trips to the New York gallery scene and participated in as many group shows that I could find. By the end of the decade I had been picked up by the Grace Harkin Gallery in the East Village. It was during this time that I participated in a group show at the New York University's 80 Washington Square East Galleries, curated by Ivan Karp of the famous OK Harris Gallery of Art in SoHo. He liked my work so much he offered me a one man show at his gallery.

Being accepted by Ivan Karp was a major achievement. He had an eye for art and an affinity for the preservation of architecture.[1] After years of salvaging the sculptural remnants of demolished buildings throughout the city in the early 1950's he founded the *Anonymous Arts Recovery Society* and eventually became one of the most famous art dealer's in New York. Karp is credited with discovering some of the most important artists of the twentieth century such as Andy Warhol,* and now George Haas. My work was finally being recognized and shown on "Broadway." I felt I was on top of the world.

Along with art, I had a great interest in archaeology. My early interests were focused on the artwork of the Maya and Aztec and Native American cultures. I attended glyph workshops at the University of Pennsylvania and learned to read Maya hieroglyphs. I also joined the Pre-Columbian Society, holding membership with the University of Pennsylvania and its affiliate in Washington, DC.

After I found a book by Randolfo Pozos titled *The Face on Mars: Evidence for a Lost Civilization?*[2] my studies quickly turned to Mars. The structural formations presented in his book were truly remarkable. Soon after I read his book I became aware of Richard C. Hoagland's UN Briefings that were offered in a video *The Moon/Mars Connection.* His presentation was so convincing that I invited family and friends to my house for a "Mars Party" to showcase the video. It was such a success that I decided to get more involved with this research and founded The Cydonia Institute in 1991. The group consisted of a small team of researchers that conducted a broad investigation of the early NASA images.[†]

In anticipation of NASA taking a new image of the famous Cydonia Face on Mars, I began posting on Richard C. Hoagland's Enterprise Mission discussion board in 1998. Soon after joining, I met geomorphologist William R. Saunders. We both enjoyed discussing our finds on and off the board and became good friends. It wasn't long before Saunders became a member of The Cydonia Institute, and we spent the next several years analyzing the early Viking archives and the newly acquired Mars Global Surveyor images of the Cydonia Face on Mars and its surrounding structures. As a result Saunders and I coauthored two books. Our first book, *The Cydonia Codex: Reflections from Mars,* was released in 2005,[3] and our second book, *The Martian Codex: More Reflections from Mars,* was released in the fall of 2009.[4]

Right after our second book was published we learned NASA had released

*Ivan C. Karp (1943–2011) began his art career in the 1960s as the director of the Leo Castelli Gallery in New York City. He opened his own gallery in SoHo in 1969 that he called the OK Harris Gallery. A gold plate list of artists he discovered includes Andy Warhol, Roy Lichtenstein, James Rosenquist, Claes Oldenburg, Jim Dine, and Cy Twombly.
†The Cydonia Institute was founded in 1991. Its original membership included George Dutton, Lee Bogart, and Pam Feather.

a new image of a Parrot Geoglyph that was prominently featured in our current book, which had been published just a month before. The new image was taken by the Mars Global Surveyor spacecraft with the highest resolution yet acquired.[5] The new image was not only amazing, but it also confirmed every avian feature that the Parrot Geoglyph possessed. This new image would have made a great addition to our book.

Over the next two years Saunders and I put all our energy into working with this new image of the Parrot Geoglyph. We were steadfast in producing a science paper that would present a convincing case that the Parrot Geoglyph was an artificial construct sitting within the Argyre Basin of Mars. We added two of our close colleagues to the paper: the guy that discovered the parrot, Wil Faust, and a longtime Mars researcher and image specialist, James S. Miller. As the paper progressed, we contacted three veterinarians to examine the parrot formation, one being an avian specialist. Once on board, the three veterinarians independently confirmed the Parrot Geoglyph had over seventeen points of anatomical correctness. Unfortunately, our first submission of the paper to a science journal was rejected. They said we needed to add another geologist.

As if by some grand design of destiny, within days of our rejection I received a Facebook message from a geologist, Michael Dale. He said he liked my work and asked if he could join The Cydonia Institute. I told him about the pending paper, and he agreed to look at it. With Dale on board, the paper was quickly accepted and published in the *Journal of Scientific Exploration* in the fall of 2011.[6] A year later the Parrot Geoglyph appeared on the front page of the *Wall Street Journal*.[7] We were all flying high.

After ten years of study and analysis of NASA and ESA images of the planet Mars and the publication of our first science paper, Saunders and I were invited to become members of the Society for Planetary SETI Research (SPSR). Once members, we had the opportunity to work alongside many of the pillars of early Martian research such as Dr. Horace Crater, Dr. Stanley McDaniel, Dr. Mark Carlotto, and Dr. John Brandenburg. Over the next six years Saunders and I produced five more papers with our SPSR colleagues; these were published in peer-reviewed science journals.

The journal publications exposed our work to a broader audience and led to invitations to appear on the History Channel's *Ancient Aliens, The Proof Is Out There,* and *The UnXplained* with William Shatner. It was during this time that Saunders and I became preoccupied with our own solo projects and began working on independent papers and books. I decided to move away from the vast assortment of half- and two-faced formations that we had collected over the past thirty years and focus on geometrically shaped structures that were easily seen. Like Ivan Karp, I also have an eye for art and architecture, and this book, *The Great Architects of Mars*, is the result of just a few of those observations.

Remote Sensing and Learning to See Past the False Image

Satellite's gone, way up to Mars
LOU REED, "SATELLITE OF LOVE"

Aerial Observations

MILITARY COMMANDERS along with artists and field archaeologists have long sought the advantage of having a bird's-eye view of the vast and expansive terrain that surrounds them. Hot-air balloons fulfilled this desire and were used for aerial reconnaissance missions by the French Aerostatic Corps during the Battle of Fleurus in 1794.[1] As hot-air balloons became more accessible to the public, an innovative English artist, Thomas Baldwin, used a balloon to produce panoramic drawings of the small town of Chester in the eighteen hundreds. He provided his patrons and local citizens with stunning views of the tranquil landscapes around them.[2]

Everything changed for aerial observations with the commercial availability of cameras in the early 1840s. A French photographer, Gaspar Félix Tournachon, took the first known aerial photograph in 1858 from a tethered hot-air balloon. He took a picture of the village of Petit-Becetre from an altitude of over 250 feet above the ground.[3] It wouldn't take long before cameras were strapped on everything from kites to blimps and even homing pigeons.

The next level of aerial observations began soon after the first successful flights conducted by the Wright brothers in 1903. Pilots would attach cameras to their airplanes, allowing them to take high-altitude pictures. This innovation quickly led to the military developing a whole new squad of aerial photographers.

The use of cameras on board airplanes not only became a crucial tool for military reconnaissance, but archaeologists also adopted the practice. By the early 1930s a pioneer in aerial photography, Professor Erich F. Schmidt, and his wife, Mary-Helen Schmidt, acquired a camera and began the Aerial Survey Expedition in 1935. The couple conducted flights over the deserts of Iran photographing the ruins of ancient cities to document many disappearing archaeological sites. The aerial images were used to establish site locations and record them on maps that future archaeologists could use.[4]

One of the first major discoveries acquired with the use of aerial photography was the mysterious Nazca Lines in Peru. These odd linear features were first observed in the 1930s, when trans-Andean aviators began flying over the arid Nazca plateau. Pilots saw a vast assortment of lines that formed pictographic images of different bird and geometric patterns scattered across this ancient landscape.[5] One of the most popular linear drawings is an image of a Condor (Fig. I.1).

Etched within the dark surface, the Condor formation has a symmetrical design that projects the silhouetted shape of an open-winged bird seen from above. A set of parallel lines form a long beak protruding from a small round head that's attached to the main body, which is oriented in a southeastern direction. A large pair of outstretched wings extends above a pair of legs with large, clawed feet. The clawed feet flank a set of splayed feathers that form a paddle-shaped tail.

Fig. I.1. Condor.
Nazca Lines, Peru.
Map data © Google.
Drawing by the
author.

Map data © Google.

Fig. I.2. Astronaut (spaceman). Nazca, Peru, 500 BCE.

Another popular image is a large figurative geoglyph known as the Astronaut or the Spaceman (Fig. I.2). Notice its facial features are extremely basic. It has a round, bulbous head with two circular eyes and a small, round mouth. The long, slender body has two legs and blocky feet that resemble the clay animation figure known as Gumby. The figure carries a large bundle on its right side and stretches its left arm up above its head.

Speculation about the origins of these line drawings and who might have produced them was offered by both the public and the scientific community. Some researchers thought that the Nazca people created them to communicate with their deities in the sky. Some thought they were reflections of constellations, while others thought they were the work of aliens and used as landing sites.

In 1939 Dr. Paul Kosok, a professor at Long Island University, was the first person to study these Peruvian line drawings publicly. Professor Kosok noticed that many of the lines and pictographic figures had astronomical alignments. He proposed that the figures were designed as astronomical markers that aligned with the sun and other celestial bodies that rose on significant dates along the horizon. He went on to declare that this ancient set of line drawings could be seen as the remains of "the largest astronomy book in the world."[6]

The use of aerial photography would again send the scientific community off its rails with the discovery of a pyramidal structure that might have only been built by some mysterious civilization that no one was aware of. The mystery started during World War II when a U.S. Army Air Corps pilot,

James Gaussman, reported seeing a large, white pyramid during a 1945 flight between India and China (Fig. I.3). He reported that the white pyramid was gigantic and had a highly reflective glimmering surface. The story was repeated by Colonel Maurice Sheahan, a Far Eastern director of the Trans World Airlines. He gave his eyewitness account to the *New York Times*, which published his story with a picture of the pyramid on March 30, 1947.[7] Over the years the story of the great White Pyramid of China was reduced to the vague mysteries of folklore and almost completely forgotten. Many thought the image was fabricated and it was nothing more than a hoax. Its validity would remain unresolved until the development of an aerial camera that could penetrate China's restricted air space.

Satellite Cameras

In October 1957 the entire world was set on edge when the Soviet Union launched the first artificial satellite known as Sputnik I.[8] The little metal sphere was only about two feet in diameter with four external radio antennae to broadcast radio pulses. Its orbital success stunned the world. It triggered a space race between the United States and the Soviet Union to achieve dominance in aerial surveillance and military reconnaissance. In 1958 President Dwight D. Eisenhower authorized the production of the nation's first photo reconnaissance satellites known as the Corona program. It was a top priority project that would be under the control of the Air Force and the CIA. Its main mission from 1960 to the early 1970s was to keep an eye on the Soviet Union and monitor restricted areas from space.[9]

This new satellite technology evolved quickly, and everything changed immensely in 1988 with the deployment of the Onyx satellite that was deployed on board the Space Shuttle Atlantis. This new high-tech satellite, which would be operated by the U.S. National Reconnaissance Office (NRO), was so advanced it could take images through fog and even in the dark. Its creators boasted that the digital images are so good they look like photographs.[10]

With the advent of home computers and the public's access to the World Wide Web, satellite images became available to everyone with the advent of Google Earth. Starting out as a struggling mapping company known as Keyhole EarthViewer, things did not improve for them until their maps were used by CNN during the 2003 invasion of Iraq.[11] It caught the attention of the U.S. government, most notably the Central Intelligence Agency, who saw the military applications of this innovative, state-of-the-art mapping program.[12]

Image courtesy of Colonel Maurice Sheahan.

Fig. I.3. White Pyramid. China, 1947.

It also caught the eye of the owners of Google, who acquired a version of it in 2003 and transformed it into Google Earth.[13] For the first time in history, this unique program provides an inquisitive public easy access to an interactive map of the entire world. It was so user friendly that it was soon adopted as an academic tool at universities that enabled researchers to view sites located in remote and inaccessible areas across the planet.

It didn't take long before someone noticed that Google Earth had released images of a restricted area in China that was home to the mysterious White Pyramid. Google Earth released a set of satellite images in 2007 that revealed a full-color, high-resolution aerial view of a large pyramid that looked very much like the White Pyramid photographed back in 1945.[14] The only problem was it was not covered with white stones, as reported, but was covered with piles of dirt (Fig. I.3). Perhaps the highly polished white stones the pilot saw were removed years earlier, or they were intentionally covered up with dirt.

To everyone's surprise, Google Earth images showed that the White Pyramid was not alone. It revealed the existence of many more of these Chinese pyramids, many more than anyone had imagined (Fig. I.4). Arranged in clusters, some are camouflaged by dirt and tall grass, while others have rows of trees growing on them. When asked about the origin of these pyramidal mounds the Chinese government has remained silent.

It wasn't until 2019 that officials of the Chinese government finally admitted that there were pyramidal formations within the Guanzhong Plains in

Map data © Google.

Fig. I.4. Pyramids. Xi'an, China.

Shaanxi Province, in the northern region of Xi'an. They also said that the pyramids, which date back to 6000 BCE, were burial mounds dedicated to members of China's royal families. One site that is in Xingping, Shaanxi Province, includes a large pyramidal formation that has been identified as the tomb of Emperor Wu of the Han dynasty (156–87 BCE). His burial mound highly resembles the famous White Pyramid. Unfortunately, the entire area is still off-limits to travelers because the Chinese government maintains that public tours and overzealous archaeologists could potentially damage the site.[15]

With Google Earth providing easy access to satellite images to the public, many people from all over the world began searching the globe for lost civilizations. In 2007 an independent researcher found over fifty geoglyphic formations within northern Kazakhstan, in Central Asia. The geoglyphs were designed in a variety of geometric shapes, including squares, giant rings, and crosses that range in size from 295 to over 1,300 feet in diameter.[16] One of the most interesting geoglyphs discovered in the area is a highly stylized, swastika-shaped formation (Fig. I.5). Its spiral design looks a lot like the plastic insert that was placed in the middle of a 45 rpm record, so it could be played on the standard turntable.

This massive collection of geoglyphic formations was originally exam-

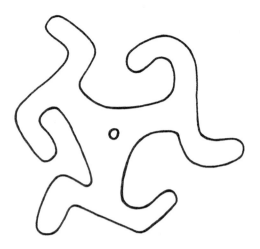

Fig. 1.5. Swastika-shaped spiral geoglyph. Kazakhstan. Map data © Google. Drawing by the author.

ined by an archaeological team from the Vilnius University in Lithuania and the Kostanay State University in Kazakhstan. The researchers used ground-penetrating radar surveys and aerial photography to expand their analysis of these newly found formations. The results of their studies were presented in a 2014 report at the European Association of Archaeologists' annual meeting in Istanbul.[17] Attempting to figure out a timeline for their construction, some of the attendees concluded that these formations were constructed anywhere between three thousand and seven thousand years ago. They also believed that they were probably the handiwork of an unknown people that occupied the area.[18] Other researchers contend they were produced during the iron age around 2,800 years ago, while some have pushed the date even further back to the Mahandzhar culture, which occupied the area around 8,000 BCE.[19]

It became quite clear that no one knew who produced these amazing formations, and although the exact origins and purpose of their creation is still being debated, scientists agree that they are extremely important. Researchers maintain that these formations are so important that they are on par with the Nazca Lines in Peru and just as important as the Great Pyramids of Egypt, and believe they should be protected and added to the UNESCO World Heritage List.[20]

In 2011 thousands of geoglyphic formations were revealed in an area spanning from Syria to Saudi Arabia that scientists and archaeologists have also compared to the Nazca Lines. The formations were discovered with the use of a satellite and aerial photography program based in Jordan. When viewed on the ground the formations are constructed from small rocks and field stones, which are arranged to produce a wide variety of geometric designs. The designs

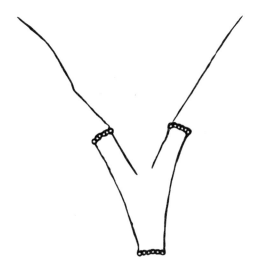

Fig. I.6. V-shaped kite geoglyph. Saudi Arabia. Map data © Google. Drawing by the author.

include kite-shaped formations, circles, triangles, and wheels (Fig. I.6). Some of the formations measure twenty-five to seventy meters across.[21]

In an effort to explain their function, the scientific community has suggested that these formations may have been used for everything from territorial markers to gates for herding animals. Others think they are burial grounds and could have been used as ceremonial sites. Some suggested they may have been used to predict astronomical alignments. Unsure of who produced these formations and what purpose they may have served, the current consensus is that they were probably produced by some unknown culture somewhere between two thousand and nine thousand years ago.[22] Again, the simple truth is they just don't know who produced any of these geoglyphs or why. It is difficult to make sense of these discoveries because they are so fresh and untested that archaeologists have not had adequate time to fully digest their meaning or their cultural impact on history.

During the same year a team of researchers led by a professor of archaeology at the University of Alabama, Sarah Parcak, released the results of their eleven-year remote sensing project of the Egyptian desert, which had amazing results. Utilizing the aid of high-resolution satellite imagery provided by NASA and a commercially available earth observation satellite known as QuickBird, the team discovered the remains of nearly three thousand ancient settlements including seventeen pyramids and one thousand tombs.[23] They discovered all this without stepping one foot in Egypt.

In response to the use of state-of-the-art imaging satellites to collect and analyze data over human observation, professor Parcak argues that although computer programs have been developed for scanning the surface of the ground for site detection, these technologies cannot replace the power of the human eye. Parcak warns:

> Computers simply do not have the same ability as human eyes have to pick out subtleties in remotely sensed images. Only the viewer will know what he or she is looking for, based on their background, and understanding of the archaeological situation. One cannot input the thousands of minor variables into computers that influence archaeologists when making choices about archaeological data. How will a computer be able to assess similar broad issues for ground surveying? As archaeologists, we can make choices regarding what information we want displayed on satellite imagery, and how we use that information to plan survey seasons. Computers cannot tell if a site or feature is present or not; they just facilitate the display of pixels. It is up to us to determine what those pixels mean.[24]

It has become increasingly clear that the public's access to satellite imagery has provided both field archaeologists and the casual researcher with an opportunity to discover "Lost Worlds" from the comforts of their own homes. The impact on the study of archaeology has been so great that the *New York Times* declared that because of this new satellite technology, "Google Earth has unlocked the gates to ancient mysteries around the world."[25]

False Images

Scientists and skeptics remind us we should remain cautious in accepting the types of facial or figurative formations we see within a random landscape or along rolling hills or even within a mountain range, because many of these oddities are nothing more than false images. These types of formations are normally viewed from the ground with the sky as a backdrop and rarely point skyward. They normally require unique lighting conditions and a particular viewing perspective to be fully recognized.

The Old Man of the Mountain, which is at Franconia Notch in the White Mountains of New Hampshire, is a common example of a false image that researchers use to show how these types of faces are created within the natural landscape (Fig. I.7). Notice the jagged profile of the old man's face only vaguely resembles a profile. The facial formation includes a pointy chin,

Fig. I.7. Old Man of the Mountain. Franconia Notch, White Mountains, New Hampshire.

a blocky nose, and a heavy brow. The profile is very basic and there isn't much facial detail.

The Old Man of the Mountain was so popular that it became an iconic monument, which was used as the state's emblem. It was also featured on license plates along with a U.S. postal stamp and a minted coin. Unfortunately, after many years of structural fatigue, the popular formation collapsed in 2003.[26]

Like the Old Man of the Mountain, most of these natural facial formations are crude or grotesque in some manner and generally comprise only an outlined silhouette with very little facial detail. They don't conform to the right size, shape, and orientation of a properly proportioned face. At best they are generic imprints of a face and project only the slightest hint of an eye, nose, and mouth. They never contain secondary features, such as an iris, nostrils, cheeks, defined lips, hair, or even ears. Despite the lack of an official reference guide providing a standard for designating an acceptable facial formation within a landscape as artificial, it can be agreed that the pattern-seeking mind needs only the simplest of features to see a face. Psychologists will argue that the mind's eye needs only the modest hint of a face, such as a triangular grouping of bumps or mounds set within a vacant landscape, to perceive a face[27] (Fig. I.8). Although the mind forms a visual projection of a facial formation, by transforming this group of mounds into a pair of eyes and a nose, we are aware that these are mounds and not a real face.

It is true that we are a pattern-seeking species. We can find hints of facial

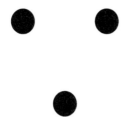

Fig. I.8. Facial projection
with three mounds.
Graphic by the author.

and figurative formations in common everyday things such as woodgrain, ink blots, and even potato chips.[28] These types of false images are commonly disproportionate or distorted in some manner and only subjectively correspond to anything remotely recognizable. We are not fooled by these false images. We know they are just natural oddities.

Simulacrum and Pareidolia

Just as scientists at the SETI Institute spend their days searching the far corners of the cosmos for digital signals that nature cannot produce,[29] I search the vast archives of NASA's Mars imagery for formations that nature cannot produce. Although the observation of unusual formations that resemble recognizable animals or face-like structures within any given landscape should be examined and challenged by secondary observers, they are too often dismissed by mainstream scientists as natural formations. They are reduced to nothing more than the brain's tendencies to find faces in rock formations by the mind creating recognizable patterns. These facial formations are thought to be the effects of our imagination or illusion-based conditions known as simulacrum or pareidolia.

The word *simulacrum* is based on a Latin word meaning likeness or similarity. It is a word often used by skeptics referring to the human mind's ability to anthropomorphize inanimate objects and for the eye to perceive facial and figurative representations in the natural environment.[30] These formations are classified as visual projections created by chance and were not intentionally created.

The origin of the word *pareidolia* finds its roots in the study of mental illness. It is a visual disorder that haunts a patient's psyche resulting in facial hallucinations as opposed to anthropomorphic projections. The word first appeared in an article written by Dr. John Sibbald in the 1868 issue of *The Journal of Mental Science*. His article reviewed a paper written by the German psychiatrist Dr. Kahlbaum titled "Delusions of the Senses," which describes a mental disorder where patients see faces everywhere around them, a disorder that Dr. Kahlbaum calls pareidolia.[31]

The word *pareidolia* was misused in the early 1990s by UFO debunker Steven Goldstein in an article published in the *Skeptical Inquirer* magazine.[32] Subsequently the word has been used to reduce any visual acknowledgment of formations such as the Face on Mars to mere projections or hallucinations. From that point on, the word *pareidolia* became politicized and quickly adopted by skeptics to discredit any facial or figurative pattern observed within a random landscape. The slanderous accusation of pareidolia is now used to convince the inquisitive public that the human eye not only seeks patterns but also can see facial features everywhere, in everyday objects.[33]

Cloud Gazing

If we look to the sky, sometimes we can see cloud formations that take on all kinds of shapes and forms. If you look long enough, clouds can begin to look like your family dog or even historical figures. Surely the young mind can imagine all kinds of things. It can see things such as the fanciful cloud formations that were perceived in a famous episode of *The Simpsons* TV show.

The segment begins as Bart and his classmates are lying on the ground cloud gazing with their bus driver, Jimbo Jones. All the kids describe seeing different things in the clouds, and at one point the bus driver says, "That one looks like a school bus going over a cliff in flames with kids inside screaming" (Fig. I.9).

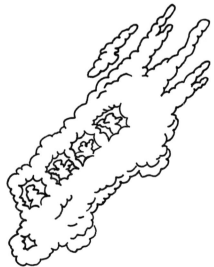

Fig. I.9. Flaming school bus-shaped cloud. Image source: *The Simpsons,* "The Telltale Head," 1990. Drawing by the author.

Image courtesy A. Joy Cole, DVM.

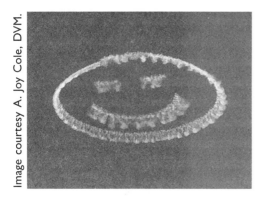

Image courtesy Rich Smith.

Image courtesy Linda Smith.

Fig. I.10. Skywriting and contrails.
Top Left: Smiley face (sky writing). Orlando, Florida.
Top Right: Triangular formation (contrail). New Mexico.
Bottom: X formation (contrail). New Mexico.

As children, I'm sure we have all had similar experiences while gazing across a clear blue sky full of oddly shaped cloud forms in the safety of our own backyards. We all realize we were looking at clouds and not artwork. A healthy, rational mind is quite able to distinguish between oddly shaped cloud forms and the straight lines of artificial contrails expelled from airplanes.

We all understand that if we see a smiley face or a triangular formation or even a large letter *X* floating in the sky above, they were not naturally formed clouds. We all know that someone produced them (Fig. I.10). As Bob Dylan once sang, "you don't need a weatherman to know which way the wind blows." It is also true that you don't need to be a meteorologist to recognize a cloud when you see one.[34]

We are all capable of recognizing the curling and wispy shape of cirrus clouds or the fluffy cotton-ball shape of cumulus clouds. We all know to seek shelter when we see the ominous approach of dark nimbus clouds. We all understand that these are natural, common cloud formations. Therefore, we should all be able to maintain the same discerning eye when looking at landforms in our surrounding landscape or on the surface of Mars. Just as we understand that contrails and sky writing are not the result of normal atmospheric conditions, when looking at the surface of Mars, I do not think we need to have a degree in geology to understand the difference between common rock formations and aesthetically designed structure.

Learning How to See

For the human eye to be able to recognize an object within a visual plane it relies on the process of "image formation" and "image fusion." Our perception of any given image is formed by our eye's desire to organize the data in its visual field in an effort to form a complete image. The eye perceives the visual field as a whole, as opposed to individual data bits of isolated elements.[35]

When viewing black-and-white images in a photograph or on a computer screen or in a newspaper, the human eye is presented with a set of pixels or dots arranged in horizontal lines and vertical rows that create a visual picture. While the human eye is capable of only seeing thirty shades of gray,[36] the digital images recorded by an orbital camera are capable of recording 256 shades of gray, which is much more information than a human eye can visually detect. While scanning this rich field of black-and-white tonalities the eye searches for visual characteristics that are darker or lighter in contrast. The eye identifies similar shapes that are grouped closer to those in the perceived viewing field, such as black shapes found within white areas giving form and shape to the image. This type of image completion, which separates contrasting groups of color, is called "relative density."[37]

Before the human eye can fuse these pixels and dots into a recognizable image it needs to be provided with a spectrum of graduated values of black, white, and gray. As the eye recognizes these basic tonalities, it also relies on the principles of depth perception. An image that may be seen quite clearly from twelve feet away will become indistinguishable if it is enlarged several times and viewed too closely. The image's integrity would become corrupted and abstracted, and the viewer would be unable to perceive its original content. This would result in the image being reduced to an indecipherable field of geometric patterns.[38] A similar distortion can easily happen when digital images are overprocessed and enlarged. They become overpixelated and reduced to a blurred field of blocky patterns and squares.

Imagine taking a black and white photograph of Leonardo da Vinci's Mona Lisa with a high-resolution camera. If your camera zooms in too close in an effort to capture her famous smile, her lips will be transformed to a splash of undefined brush work. However, if you adjust the focus and calculate the image fusion to the proper viewing distance, her lips will take on their intended form and you will once again see a smiling mouth with slightly parted lips. Establishing a proper viewing distance is also important when viewing land art, such as the Nazca land drawings in Peru. When these carved linear groves are viewed from the ground, they appear to be only random marks throughout the landscape. However, when viewed farther away from a hilltop the giant figures of animals and geometric designs begin to come into focus. When viewed from high above the ground with the aid of an aircraft all of these animals and geometric designs become fully recognizable.

In order to produce the proper image fusion of any given image, it is important to designate the distance at which the image or structure was intended to be viewed. The same standards can be applied when observing aesthetically designed landforms on Mars. However, on Mars the intended viewing distance of structures and landforms remains ambiguous and may vary depending upon the size of the formation or which portion of a formation is being viewed.

Scale is also a major variant when trying to form a cohesive image of unknown scale. Assuming these geoglyphic or geometrically designed formations on Mars were intentionally designed works of art, we must determine their proper viewing distance. Challenged with a similar task in determining the origin and purpose of artificial structures observed in orbital satellite imagery of Mars, Dr. Tom Van Flandern compiled a workable hypothesis. He theorized that any facial configuration that was designed to be seen from high above the surface, such as the Face on Mars, would have to be constructed to such a scale that it could only be adequately recognized as an intended work of art when it was observed at the proper viewing distance.[39] With these basic optical parameters in mind, he speculates that the natural viewing station for the Face on Mars, which is a mile and a half in length and a mile wide, would be from an orbiting spacecraft or from one of the moons orbiting the planet.

The Artificial Origin Hypothesis

Theoretical plasma physicist and member of the Society for Planetary SETI Research John E. Brandenburg has compiled a set of guidelines for an Artificial

Origins Hypothesis that offers four scenarios for the existence of artificial structures on Mars.

The first is the Cydonia Hypothesis. This is a scenario that suggests that the formations observed at Cydonia, such as the Face on Mars and its surrounding formations, were built by an Indigenous civilization that no longer exists. The second explanation offered is the Previous Lost Civilization Hypothesis. This model suggests that a previous highly technological civilization that once existed on Earth traveled to Mars in the distant past and was responsible for building these formations. The third scenario is the Colonization Hypothesis. This scenario suggests that the formations observed at Cydonia were produced by someone outside our solar system. A group of explorers came to Mars sometime in the distant past and built these formations for some unknown reason, and after a period of time, they left the planet. The fourth scenario is the Exploding Planet Hypothesis. This idea suggests that Mars was once a moon of a now destroyed planet and the remains of that planet now occupy the area that we know as the asteroid belt. He suggests that it was either this destroyed planet, or one of its lost moons that was the original home of these ancient builders.[40]

Presented with these four scenarios, I have attempted to establish an acceptable methodology that would support the idea that these anomalous surface features on Mars are artificial constructs that were intentionally built as geoglyphic or geometrically designed structures. In reviewing any set of anomalous formations as candidates for artificiality, we would need to develop a testable standard of evidence.

Diagnostic Criteria

The formulation and testability of any proposed hypotheses that suggest the probability of an anomalous formation on the surface of Mars should utilize the standards of the scientific method. When reviewing an anomalous formation, the proposed candidate should be supported by a controlled data set that can be observed and tested by a repeatable set of experiments.[41] Terrestrial structures can be identified by their strong geometric shapes and symmetry, while geoglyphs can be more ambiguous. Geoglyphs are produced in a wide variety of aesthetic styles that include everything from photographic perfection to the most rudimentary forms of gestural communication. The following is a set of diagnostic criteria that I have compiled in an effort to identify an artificial formation on another planet.

Class A—Descriptive Forms

Includes any formation identified with a humanoid or animal shape that exhibits pictographic, biomorphic, anthropomorphic, or zoomorphic representations.

Level 1: Silhouette

> Includes any formation identified with a gestural or generic outline that follows the contours of a profiled head or physical body.

Level 2: Basic

> a. Includes any formation that identifies a facial portrait, which includes an eye, nose, mouth, forehead, and chin.
> b. Includes any figurative formation that identifies a body that includes a head, torso, and limbs.

Level 3: Defined

> a. Includes any formation that identifies a portrait that includes an eye, nose, mouth, forehead, and chin and secondary features such as an iris, eyebrow, ears, nostrils, lips, and teeth.
> b. Includes any figurative formation that includes secondary anatomical features such as feet, hands, and digits.

Level 4: Complete

> Includes any portrait or figurative formation that includes physical features that are measurable and adhere to the proper size, shape, and orientation of accepted anatomy.

Class B—Linear Forms

Includes any wall of heaped material or cleared groove impression that creates a continuous formation or mark.

Level 1: Straight Lines

> Includes any wall of heaped material or cleared groove impression that extends out in a straight direction.

Level 2: Meandering Lines

> Includes any wall of heaped material or cleared groove impression that traverses along the surface forming zip-zag or spiral patterns.

Class C—Geometric Forms

Includes any elevated formation that includes a circular or polygonal shape. Polygons may have straight lines and angles that exhibit a high degree of regularity and symmetry.

Level 1: Circular Forms

Includes any round formation with an outer rim that has no straight edges or corners, while its contours maintain the same distance from its central point. The circular form (excluding craters) can include ovals and crescents.

Level 2: Polygonal Forms

Includes any formation with at least three straight, linear sides and angles that create triangles, squares, rectangles, and trapezoids.

Class D—Graphic Features

Includes identifiable features on a formation that has an emblematic motif or iconography that project a recognizable design. Features may include decorative ornamentations or symbols that exhibit cultural significance.

Class E—Alignments

Includes any set of formations that show a direct relationship between two or more formations that share measurable sight lines between structural points or display an orientation with due north or exhibit solar and celestial alignments.

Class F—Common Footprint

Includes any set of formations that share a direct correlation with the size and shape of its neighboring formations by exhibiting a common design or share and exhibit a common set of dimensions that include length and width.

Class G—Secondary Confirmation

Includes any formation that has unique features that can be clearly identified and confirmed by viewing secondary images. Each candidate must withstand a testable evaluation that can be repeated. The formations external shape and internal topography must be confirmed by secondary images taken at similar or higher resolution during alternate times of day and/or different seasons.

Diagnostic Evaluation

Although presented with testable and supportive criteria, applying this comprehensive evaluation to rudimentary or loosely rendered geoglyphs that are observed on Earth or on another planetary body may prove to be difficult.

Rudimentary Geoglyphs

Contemporary artists that produce environmental land art have shown that they are quite capable of producing gigantic geoglyphic images with a high level of photographic accuracy. As an example, here is an eleven-acre portrait of a young girl's face produced in 2013 by Cuban American artist Jorge Rodríguez-Gerada. The portrait, titled *Wish*, was produced with the aid of state-of-the-art Topcon GPS technology.[42] It was created with the help of local volunteers that utilized an estimated thirty thousand pegs, two thousand tons of soil, and two thousand tons of sand.[43] Although the clarity of this portrait is quite remarkable and photographic, it is quite unique. Most and possibly all the geoglyphic formations produced throughout the world by ancient cultures were rendered in a much simpler and rudimentary manner.

In 2018 a team of archaeologists reported that they used commercially available drones to discover an entire set of unknown geoglyphic line drawings within the Pampas de Palpa region of Peru. The formations are near the border of the famous Nazca Lines, and scientists speculate that they might

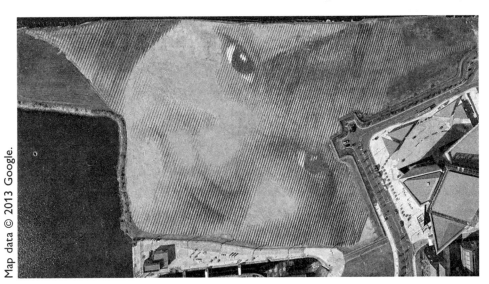

Fig. I.11. *Wish* by Jorge Rodríguez-Gerada. Belfast, Northern Ireland.

Map data © 2013 Google.

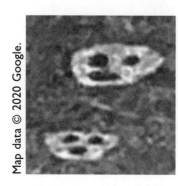

Map data © 2020 Google.

Fig. I.12. Head geoglyphs.
Pampas de Palpa, Peru,
circa 500 BCE.

predate the Nazca Lines, which were produced somewhere between 500 BCE and 200 CE.[44] The new geoglyphs feature extremely simplistic renderings of various anthropomorphic and zoomorphic figures and at times are just isolated heads. Here are two examples of the types of heads that have been identified with rudimentary facial features that are found etched along a sloping hillside (Fig. I.12). The two heads are simply composed of three dots, set within a triangular pattern, inside of a circle. The dots are intended to represent a pair of eyes and a nose. Their simplistic, cartoonish design looks very much like the three mounds illustrated in Fig. I.8 that project the basic form of a face. If these two heads were found on Mars, would scientists dismiss them as natural facial projections?

Exploring an area further down the South American coast, archaeologists encountered another geoglyph that questioned the aesthetic standards of the scientific community. During the excavation of the ancient ruins of Caral, Peru, a site that dates to well before 2600 BCE, a partial portrait of a human head was discovered in early 2000 and was featured in *Smithsonian Magazine*[45] (Fig. I.13). The rudimentary image is composed of a linear formation which has a childlike simplicity that only captures the basic form of a human head. It was produced by precisely placing various sizes of stones across the surface of the barren desert.[46] Notice how the D-shaped head is created with a sweeping mat of raked hair and a large, gaping mouth. The forehead appears incomplete, and there is no evidence of an ear or neckline. Its facial features include a large nose and a small, undefined, football-shaped eye.*

Another example of a rudimentary composed geoglyphic formation was discovered in 2012 in the Ural Mountains of the Chelyabinsk region of

---

*For some unknown reason the image of the Half-Face geoglyph was removed from the Smithsonian's website article sometime in 2005.

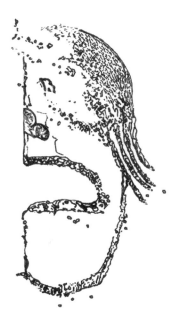

Fig. I.13. Half-Face geoglyph. Caral, Peru. Drawing by the author.

Russia (Fig. I.14). Estimated to be over three to four thousand years old, this childlike drawing of a large elk or moose was noticed by regional specialist Alexander Stanislav after studying aerial photos of the area that were posted on the internet. Measuring 715 feet long and 900 feet wide, the outline of the elk-shaped formation was created by digging a linear trench that was filled in with a variety of large and small stones.[47] The formation is not anatomically correct and appears as an awkwardly rendered, gestural drawing. It has a long, thick muzzle and a set of asymmetrical antlers. One antler appears as a thin tubelike horn, while the other is thick and jagged like a flattened

Fig. I.14. Elk. Ural Mountains, Russia, circa 2000 BCE. Drawing by the author.

moose's horn. Its body comprises an arching back supported by four legs that vary in size and length.

According to the diagnostic criteria that I have set forth, to establish an acceptable criterion for verifying artificial formations on another planet, skeptics would quickly dismiss these three geoglyphs as natural formations and set them aside as geological oddities that only slightly resemble a human face or an animal.

Unresolved Structural and Geoglyphic Anomalies

J. P. Levasseur, a physicist, and member of the Society for Planetary SETI Research reminds us that when an unusual and unexplained formation is observed, outside of the accepted realm, the first thing critics demand is to know how and why such a formation was produced. He warns that any causal mechanisms should be worked out, only after the anomaly is first observed and thoroughly studied. He cautions that "All scientific evidence is based on observations of some kind, and it is only after we observe and study the available evidence that we are able to develop a causal mechanism. It is not our burden to establish a viable and believable origin scenario to justify the existence of these formations, because that would be like putting the cart before the horse. In that scenario, you nip science in the bud. We must investigate these formations through careful observations of the available data set and if multiple images support artificial origins—it is only then that we would be able to speculate on how they got there."[48]

American planetary scientist, Carl Sagan, popularized a "hard line" standard for accepting the discovery of artificial structures on another planet such as Mars. Known as the "Sagan standard," he declared "extraordinary claims require extraordinary proof."[49] Hence forth, this has become the high bar of scientific affirmation provided by skeptics for their acceptance of any planetary anomaly. The skeptic assumes that every scientific discovery ever made throughout history was held to this extreme standard of proof. They believe that no new discovery can ever be trusted or believed without gaining the consideration of the scientific community and becoming an accepted science. This surely is not true. One needs only to look to the many unexplained archaeological sites and geoglyphic formations that exist right here on earth, to find fault in this extreme standard.

There are many structures around the world that defy logic and man's technological abilities. Just look at the Great Pyramid at Giza in Egypt and the megalithic ruins of Puma Punku, a small site that is part of the Tiwanaku in western Bolivia. They are perfect examples of unsolved, technological anomalies. These are just two enigmatic sites that have been intently studied for decades and decades, while their construction and origins are still puzzling to scientists and attract strong debate. All one has to do is examine the complexity of the interior chambers and passageways constructed within the Great Pyramid at Giza to get a highly volatile discussion started (Fig. I.15).

In November 2017 the journal *Nature* published a paper reporting that a group of researchers had discovered a large void inside the Great Pyramid at Giza.[50] The mysterious chamber was discovered just above the Grand Gallery by an international team of experts after they utilized a revolutionary imaging technique that allowed them to peer through the thick walls of this ancient Egyptian pyramid. They harnessed subatomic particles called muons, which are by-products of cosmic rays that are formed in the Earth's upper atmosphere, to look through the stone walls. The imaging technique is similar to the way doctors use X-rays to see through flesh. After reviewing the published report Egyptologists, along with fellow archaeologists and scientists, were stunned. They were all left clueless as to what this chamber was used for.[51]

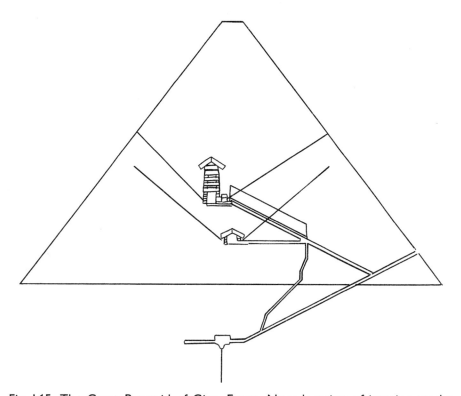

Fig. I.15. The Great Pyramid of Giza, Egypt. Note location of interior tombs, passageways, and air shafts, known prior to the discovery of the void. Drawing by the author.

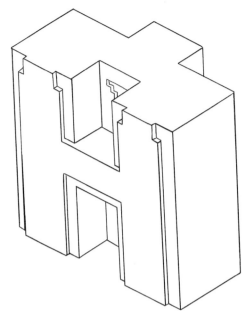

Fig. I.16. H-block. Pumapunku, Bolivia.
Drawing by the author.

A similar debate has evolved on the other side of the world, where archaeologists and expert engineers are confronted with dozens of cookie-cutter-shaped "H-blocks" found scattered around Pumapunku (Fig. I.16). After examining their flat, mirrorlike surface and precision-cut artisanship, many researchers are left scratching their heads. No one knows how these blocks were carved or how they achieved such exquisite execution of high-tech "machine tooling."[52] No one really knows why they were produced or what they were used for.

Massive structures are not the only problem. One of the most unresolved and baffling geoglyphic anomalies was discovered by charter pilot Trevor Wright back in 1998, as he flew over the Finnis Springs plateau in Southern Australia.[53] It was there that he observed the remains of the mysterious figure carved in the surface that has become known as the Marree Man (Fig. I.17). Named after the nearby town of Marree, the two-and-a-half-mile-long geoglyph has attracted worldwide attention from both the public and the scientific community ever since its discovery. Right from the beginning its origins and its true identity have been pondered and debated and constantly misinterpreted. It was first thought to represent everything from the Greek god Zeus wielding a lightning bolt[54] to an Egyptian representation of their bird-headed God, Thoth. These early interpretations prompted many of the researchers to

Fig. I.17. Marree Man.
Finnis Springs Plateau,
Australia.
Map data © Google.
Drawing by the author.

entertain the idea that there must have been some long-lost Greek or Egyptian settlement that once visited this exotic land down under.[55]

To determine its origins a meticulous examination and careful reconstruction of the formation's contours was attempted in 2016.[56] As a result most observers now agree that the formation actually depicts a bearded, indigenous hunter and not a Greek god or bird-headed Egyptian. The bearded figure stands naked, not wielding a lightning bolt, but holding a stick or boomerang, two deadly weapons that were used by aboriginals to kill birds. The premature "bird head" interpretation was fostered by the shape of the hunter's bundled headdress that protrudes out from the back of his head like a beak.

Over the years various theories and stories have evolved attributing the construction of the massive figure to everything from the indigenous aboriginal people to the engineering skills of ancient aliens. Some believe it was produced by the Australian Army or the US Air Force that maintained the Joint Defense Facility that was located nearby.[57]

Others contend the formation was produced by a secretive group in honor of the famous Outback explorer John McDouall Stuart, who visited the area in 1859,[58] thereby labeling the formation Stuart's Giant. There are even fringe groups that believe it was produced by a radical sect of the Branch Davidians. The most popular theory attributes the work to a local artist known as Bardius

Goldberg, who died in 2002. The only problem with this story is that there is no record of the artist ever admitting or denying that the work was his.[59]

In 2018, during the twentieth anniversary of its discovery, the search for answers intensified. So much so that an Australian businessman and philanthropist, Dick Smith, revealed that "after over two years of very thorough research, I have no evidence whatsoever who did it."[60] In an effort to put an end to this outback enigma, Smith went public and offered a $5,000 (AUD) reward to anyone who could solve this enduring mystery.[61] As of the date of this printing there have been no takers.

In the end the formation's origins are left unresolved and the scientific community is still baffled by its mere existence. As a result, the Australian people are left to believe whatever story they chose, and the world is still wondering who produced this giant figure. All we are sure of is that it exists.

The Eye in the Sky

Humans have a long history of altering their environment by producing an extensive lexicon of geometric and pictographic earthworks. Archaeologists believe that these early formations were created by some of our earliest cultures to establish memorials or monuments for worship and sacred rituals. Astronomers speculate that many of these mounds and linear formations may have been created to represent prominent constellations or to mark important planetary and solar alignments.

The creation of geometrically shaped mounds and geoglyphic art works may also have been produced as territorial markers. They could have been produced to establish tribal boundaries that could be seen from a high vantage point, such as a surrounding hillside or a distant mountain peak. Still, others believe they were constructed for no other reason than to communicate with the gods above or be seen by the watchful eye of extraterrestrials.

In the 1820s Carl Friedrich Gauss, a well-known German physicist and mathematician, had the idea of creating an immense geometric landform to communicate with extraterrestrials. He proposed the construction of an enormous diagram depicting the Pythagorean theorem, also known as the 47th Problem of Euclid, in the thick Siberian forest (Fig. I.18).

The proposed landform would consist of one large right triangle and three squares cut into the dense pine forest. Once the imprint was complete, wheat would be planted inside each of the cleared areas to provide a contrasting color to the pine trees. This massive agricultural imprint would be so large it could

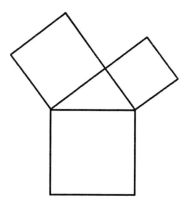

Fig. I.18. Pythagorean Theorem. Diagram proposed by Carl Friedrich Gauss, circa 1820. Graphic drawing by the author.

be seen from the Moon or Mars. Gauss believed that a complex geometric image of the Pythagorean Theorem would demonstrate the existence of intelligent life on Earth and get the attention of alien observers.[62] His proposal was never realized.

In 1947 a Japanese American sculptor, Isamu Noguchi, envisioned a similar idea. He proposed the construction of a large-scale sculpture of a human face that could be seen from the surface of Mars. His design incorporated a grouping of geometrically shaped earthen mounds to create the most basic facial features of a human face (Fig. I.19). Inspired by the postwar threat of nuclear war and global annihilation, his proposed geoglyph *Sculpture to Be Seen from Mars* would be well over a mile long and hopefully survive as a lasting memorial to the human race.[63]

Fig. I.19. Proposed model for *Sculpture to Be Seen from Mars*, 1947. Drawing by the author after Isamu Noguchi.

Whatever rationale we use to consider or reject the idea of constructing such enormous geometric and geoglyphic formations here on Earth, it is becoming clear that mankind's obsession with transforming his environment and producing pictographic or geometric monuments is a long-held human tradition. Perhaps Gauss and Noguchi were not the only visionaries to have contemplated the idea of constructing a visual "marker" that could be seen from space by a watchful eye in the sky and establish contact between two worlds.

This very question of finding a "marker" on another planet was addressed by a group of mainstream scientists in a book that was released in 2014 entitled *Archaeology, Anthropology, and Interstellar Communication*.[64] The report, which was led by astrobiologist Douglas A. Vakoch, included NASA and SETI scientists, along with archaeologists and anthropologists, that determined the observation of rock art and sculptural carvings on a planetary surface should be considered as possible examples of extraterrestrial communication. The authors make the case that scientists may have difficulty identifying "manifestations of extraterrestrial intelligence" because they might "resemble a naturally occurring phenomenon." This leaves the door open for the idea that an unknown, lost civilization could have left us a message on Earth or on our moon or even on Mars that we are totally unequipped to understand or even recognize.[65] We might want to rethink the idea "if you build it, they will come."

If You Build It, They Will Come

After many years spent studying the anomalous surface features observed on Mars, plasma physicist Dr. John E. Brandenburg has concluded that there is much more to these formations than first meets the eye. He believes the mere existence of these pictographic and geometrically shaped structures provides enough evidence to suggest that Mars was an occupied planet. He hypothesizes that Mars once had an Earthlike climate with oceans and was home to both plants and animal life, including an advanced humanoid civilization. He envisions them as a Bronze Age mound-building culture similar to those that arose on Earth. Then at the height of their technological advancements this civilization was quickly wiped out by a more advanced aggressor from another planet.[66]

His hypothesis is expanded in his 2015 book *Death on Mars*. In it he offers visual and scientific data supplied by NASA to reveal the existence of a dead civilization on the red planet. Since the early Viking missions NASA has not only detected the remains of artificial structures on the surface of Mars, but they have also detected the remains of high levels of Xenon 129 within its atmosphere. A chemical isotope, Xenon 129 is a nuclear signature. It indicates that the planet Mars was the victim of a massive thermonuclear explosion sometime in its distant past.[67] According to Brandenburg two above-ground nuclear bombs were exploded over the Cydonia and Utopia regions, destroying this once thriving culture.

To make sense of this massive destruction, Brandenburg suggests it might be explained by the Fermi paradox. The paradox is used to explain the alien contact dilemma, a situation where scientists tell us that the universe is brimming with planets suitable for life, however, we only hear silence. When the famous physicist Enrico Fermi (known as "the architect of the atomic bomb") was confronted with this problem, he responded, "Where is everybody?" hence the Fermi paradox. Brandenburg also asks us "Where is everybody?" He warns that an aggressive alien civilization might have discovered a young, thriving culture living on Mars and destroyed them.

The Martians may have invited their own destruction by announcing their presence. It might be a good idea for any advanced, highly technological civilization to remain silent and keep their existence under the radar. Contacting ET could be deadly. The lesson of Martians building architectural beacons may be "If you build it, they will come." And when they come, they may not be friendly. As a precaution Brandenburg has invited NASA to send astronauts to Mars immediately and find out what happened to the Martians and learn how to avoid their fate.[68]

Theoretical physicist Stephen Hawking has also warned against seeking aliens. He believes that we should not respond to any signals from some far-off planet because it might attract unwanted attention. In his film, *Stephen Hawking's Favorite Places*, he warns that contacting an advanced civilization could be devastating. He compares any such contact with aliens to the time Native Americans first encountered Christopher Columbus, and as we know things did not turn out so well for them.[69]

All we know about Mars is that at one time in the past it was a lavish Earthlike planet that supported life, and then as everything was going well, something happened, and it was destroyed. Heeding Hawking's and Brandenburg's warnings and setting aside the visionary beacons proposed by Gauss and Noguchi, perhaps a comprehensive study of the anomalous formations that have been observed on Mars will answer a formidable set of questions. Is our heritage vastly older than the dirt that has been used to cover it up? Does our point of origin somehow begin with the rise and collapse of the red planet, Mars? Who built these structures on Mars, and why were they abandoned? Were our ancestors the Martians, or are we connected to some unknown civilization that once visited or even conquered Mars? And if so, did they leave us a familiar and recognizable set of formations on its surface as a beacon to attract our attention and recapture our hidden legacy?

If so, I believe I am well armed with an immense arsenal of comparative formations in the pages that follow. And even though we don't know who built many of our own terrestrial formations, and we don't fully understand how or why many of them were constructed, we do know that they exist, and someone built them. Perhaps the pictographic and highly symmetrical, geometric formations discovered throughout the red planet can be given the same consideration and offered a fair and open scientific review. I believe that through NASA's own images the truth will be revealed.

ONE

The Sagan Pyramid

Mariner 9

ON MAY 30, 1971, the Mariner 9 spacecraft was launched from Cape Canaveral Air Force Station in Florida and reached the planet Mars on November 14, making it the first spacecraft to orbit another planet. Its mission was to conduct atmospheric studies and map the Martian surface. The onboard camera was capable of acquiring images from the lowest altitude, with the highest resolution yet achieved. Pictures were taken from an altitude of 930 miles above the surface with a resolution between 1,100 and 110 yards per pixel.[1]

After months of delays caused by annoying dust storms the spacecraft was eventually able to map 85 percent of the Martian surface and send back more than seven thousand pictures.[2] Mariner 9 revealed evidence of ancient riverbeds and the remains of a gigantic canyon over 2,500 miles long called Valles Marineras. It also acquired pictures of the polar caps and discovered Olympus Mons, the largest known volcano in the solar system. The two small moons of Mars, Phobos and Deimos, were also photographed.[3]

As scientists began examining the new images of Martian geology, one image taken of the Elysium area in 1972 appeared to include a set of anomalous formations (Fig. 1.1) that caught the attention of a pair of geologists following the mission. In 1974 an article written by Mack Gipson Jr. and Victor K. Ablordeppey titled "Pyramidal Structures on Mars" appeared in the scientific journal *Icarus*. Their article reported the following:

Triangular and polygonal pyramid like structures have been observed on the Martian surface. Located in the east central portion of Elysium Quadrangle (MC-15), these features are visible on the Mariner 9 photographs B frames MTVS 4205-3 DAS 07794853 and MTVS 4296-24 DAS 12985882. The structures cast triangular and polygonal shadows. Steep-sided volcanic cones and impact craters occur only a few kilometers away. The mean diameter of the triangular pyramidal structures at the base is approximately 3.0 km, and the mean diameter of the polygonal structures is approximately 6.0 km.[4]

The same pyramidal structures were also noticed by world-renowned astronomer Carl Sagan. In 1977 Sagan was invited to speak at the Royal Institution in London where he presented six lectures on the solar system in a series called The Planets. In one of the lectures titled "Mars before Viking" he introduced an audience of school children to an image of a pyramidal formation found on Mars that was captured by the Mariner 9 spacecraft in 1972 (Fig. 1.1). He speculated that the pyramidal formations may have been created by the effects of high winds and harsh sand blasting that transformed the large mounds into these pyramidal shapes. At the close of his lecture, he left the audience with a closing thought in which he not only acknowledged that these formations were indeed "lovely sculpted objects" but we needed to know what they actually are.[5]

Sagan was so intrigued by this pyramidal formation that he included it in his 1980 book and television series, Cosmos. In his book he made the following comment.

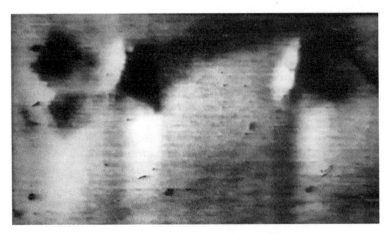

Fig. 1.1. Elysium Pyramids, Mars.
Detail Mariner 9 image 4205–78 (1972).

The largest Mars Pyramids have a base width of 3km and a height of 1km, so they are much larger than the Pyramids of Sumer, Egypt and Mexico. With the ancient, eroded shape, they could be small hills, sandblasted for centuries, but they need to be viewed from nearby.[6]

Elysium Planitia

The Elysium Planitia region of Mars is located in the eastern quadrant of the planet, just above Terra Cimmeria and below Utopia Planitia. The area expands across a broad, windswept plain filled with oddly shaped mounds and knobs which straddles the equator at approximately 3.0°N to 40.0°N and extends from 154.7°E to 180.0°E.[7]

It is a volcanic region that contains major volcanoes such as Elysium Mons and Albor Tholus. Recent data suggests that volcanic activity may have occurred as recently as fifty-three thousand years ago, creating an environment that was suitable for supporting life.[8] Water has also left its mark on the region in the form of riverbeds and canyons. Scientists contend that the geology of the area, which includes marine sediments created by glaciofluvial activity, provides evidence that water played a substantial role in affecting this region during the recent past.[9]

In March 2017 NASA scientists announced that the flat planes of Elysium Planitia would be ideal as a landing site for their upcoming robotic lander InSight. The lander was launched on May 5, 2018, from Vandenberg Air Force Base in California and successfully landed on Mars on November 26, 2018.

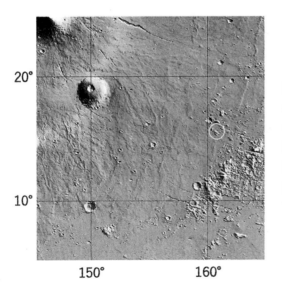

Fig. 1.2. Elysium Planitia region of Mars. Mola map, notated with the location of the Sagan Pyramid (circled).
Image courtesy NASA/JPL/ Malin Space Science Systems/ The Cydonia Institute.

The InSight lander was sent not only to study marine sediments but to study the planet's internal heat and seismic activity.[10]

The word Elysium comes to us from the stories of Greek mythology. It is a reference to a special place in the afterlife, known as the Elysian Fields. It was a place reserved for virtuous men that were considered heroes and close friends of the gods.[11] The area of this study is located at around 15°N and 161°E, just to the southeast of the extinct volcano Albor Tholus (Fig. 1.2 on page 33).

Viking

The Viking 1 spacecraft was launched atop Titan IIIE rockets on August 20, 1975, and its companion craft, Viking 2, was launched a month later on September 9, 1975. Viking 1 began orbiting Mars on June 19, 1976, with Viking 2 entering the planet's orbit on August 7. The Viking camera was able to acquire images of the surface at a resolution of 150 to 300 pixels per meter.[12] Over the course of their missions between 1976 and 1980, the Viking I and Viking II orbiters obtained thousands of images of the Martian surface that covered the entire planet.[13]

In November 1978, a year after Sagan's appearance at the Royal Institution in London, the Viking 1 orbiter had the opportunity to rephotograph the same Elysium pyramids that he mentioned in his talk. A comparison of the Viking 1 and Mariner 9 images is offered in Fig. 1.3.

The Viking image (883A03) was taken in the late morning at a lower resolution, measuring at 156 pixels per meter. The Viking image is very grainy and

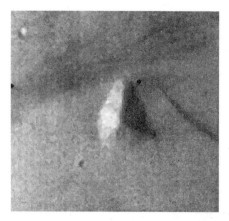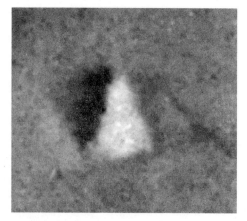

Fig. 1.3. Sagan's Elysium pyramid.
Left: Detail Mariner 9 image 4205-78 (1972).
Right: Detail Viking image 883A03 (1978).

not as sharp as the Mariner 9 image; however, you can still identify the pyramidal shape of the formation (Fig. 1.3). In his 1997 book *The Demon-Haunted World,* Sagan would again make references to the pyramids of Elysium.

> There is something a little eerie about these pyramids in the desert, so reminiscent of Egypt, and I would love to examine them more closely.[14]

Unfortunately, this second image of the area was a disappointment and left the overall morphology of the pyramidal formations unresolved for the remainder of Sagan's lifetime.

European Space Agency

Five years after Sagan's death, the European Space Agency (ESA) launched the Mars Express spacecraft on June 2, 2003, from Baikonur Cosmodrome in Kazakhstan. It reached Mars during that September and began taking images of the planet's surface the following year, in January 2004.[15] Four years later the spacecraft's High-Resolution Stereo Camera acquired a new image of the Elysium region of Mars that included the area surrounding Sagan's Pyramid.

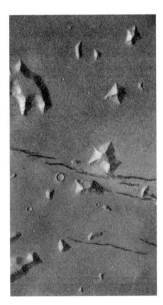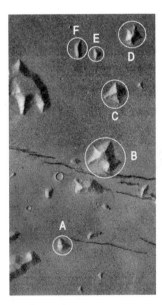

Fig. 1.4. The Sagan Complex.
Left: Expansive view of the area above Sagan's Pyramid.
Detail ESA H5208_0000_ND3 (2008).
Right: Enhancement and Notations A–F by the author.

The ESA image (H5208_0000_ND3) was taken in the early evening with a resolution of around thirteen meters per pixel.

The expansive ESA image provides a broad view of the area around Sagan's anomalous pyramid and reveals for the first time that his pyramid is not alone. We can now see that Sagan's pyramid is surrounded by a variety of geometric and pyramidal formations that demand further analysis. In Fig. 1.4 on page 35 I provide a site map of the Sagan Complex in which I have labeled some of the most interesting formations A through F. The Sagan Pyramid is labeled A.

Mars Reconnaissance Orbiter

On August 12, 2005, NASA launched the Mars Reconnaissance Orbiter (MRO) spacecraft from the Cape Canaveral Air Force Station in Florida. The spacecraft reached Mars on March 10, 2006, and began photographing the entire planet during that November. The spacecraft is equipped with two onboard cameras. The first is the Context Camera, which is called the CTX camera. This smaller camera provides images of the Martian terrain with an average resolution of 6 meters per pixel. The second camera is the High-Resolution Imaging Science Experiment, which is called the HiRISE camera. This second camera is capable of producing images with the highest resolution ever achieved, coming in at 1 meter per pixel.[16]

In September of 2007 the Mars Reconnaissance Orbiter (MRO) space-craft's HiRISE CTX camera took an image of the Elysium area of Mars that included Sagan's famous pyramid. The CTX image shown in Fig. 1.5 was taken in the early morning with a resolution of 5.5 meters per pixel. Notice the low

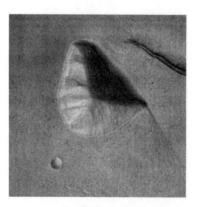
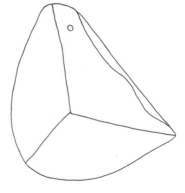

Fig. 1.5. Sagan's Pyramid (labeled A in Fig. 1.4).
Left: Detail MRO HiRISE CTX P11_005219_1961_XN_16N198W (2007).
Right: Analytical drawing by the author.

rays of the morning sun, coming in from the west, causing the eastern face of the three-sided pyramid to be in shadow. Also notice the vertical striations that run down the western side of the pyramidal formation. The formation measures 2.63 miles in width, from its western side to its easternmost point and is 2.43 miles in length from its northernmost point to its southern side.[17]

Thermal Emission Imaging System (THEMIS)

On April 7, 2001, NASA launched the new Mars Odyssey spacecraft from Cape Canaveral Air Force Station in Florida and by the following October it began orbiting the planet Mars. The spacecraft is equipped with an onboard camera called the Thermal Emission Imaging System Visible Camera (THEMIS-VIS) that is designed to provide systematic global coverage of Mars in both the visible and IR bands.

The THEMIS infrared camera on board the Mars Odyssey spacecraft records temperature changes on the surface associated with volcanic heating and near-surface water or ice. THEMIS infrared images taken during the day will look much like a shaded relief map, with areas facing the sun being bright (hot) and shaded areas being dark (cold). In a THEMIS inferred image taken at night the instrument can detect temperature differences due to the various materials present within the surface.[18]

In addition to obtaining detailed coverage of potential landing sites for NASA's planned 2003 rover mission, the primary focus of the THEMIS investigation was to collect IR data and black-and-white images. Color images would be selective and controlled by the limitations in data rates and by NASA's desire to focus on the IR data and obtaining global coverage. According to the Principal Investigator for the Mars Odyssey THEMIS spacecraft, Philip Christensen, the goal of the mission was to study the planet's "remarkable geomorphology."[19]

In 2015 the Mars Odyssey THEMIS camera acquired a beautiful image of Sagan's Pyramid in November. The image was taken with a resolution of 73 meters per pixel. The image was taken late in the day, with the light coming in from the east, resulting in a long shadow being cast across its western side (Fig. 1.6 on page 38). The triangular peak of the shadow confirms the pyramidal shape of the formation.

The new THEMIS image is about twice the resolution as the earlier Viking image. Notice the three triangular faces of the pyramid, with the darkest side on the western face and the lightest side on the northeastern face. The extended

Fig. 1.6. Sagan's Pyramid (labeled A in Fig. 1.4). Mars Odyssey THEMIS V61919005 (2015).

shadow provides a hint at the height of this enormous pyramid. Combining that with the light and dark contrast of the faces, the THEMIS image provides a real three-dimensional form to this structure. Dr. Sagan would definitely be intrigued.

Three-Sided Pyramid

Located on the opposite side of the complex is another three-sided, triangular pyramid (labeled E in Fig. 1.4) that is very similar to the original Sagan Pyramid. A higher resolution image was obtained by the MRO HiRISE CTX camera in 2007 (Fig. 1.7). The image was taken in the winter during early morning with a resolution of 5.5 meters per pixel. The image shows a triangular formation with a very smooth and highly reflective surface. It has three

Fig. 1.7. Three-sided pyramid (labeled E in Fig. 1.4). Detail MRO HiRISE CTX P03_002318_1961_ XN_16N198W (2007).

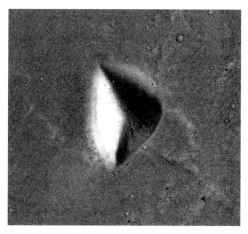 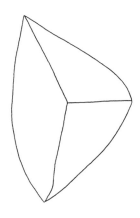

Fig. 1.8 Three-sided pyramid (labeled E in Fig. 1.4).
Left: Detail MRO HiRISE CTX P13_006142_1964_XN_16N198W (2007).
Right: Analytical drawing by the author.

sharp ridge lines that descend from a common peak creating three faces. The north and southeastern faces are of similar size, while the western face is almost twice as large.

Ten months later, a second MRO HiRISE CTX image of the area was acquired at the end of 2007 (Fig. 1.8). This second image was also taken in the morning hours during the cold winter and with a similar resolution of 5.5 pixels per meter. The formation measures 2.18 miles from its northern most apex to its lower, southern point and is 1.50 miles wide.[20] An analytical drawing is provided in Fig. 1.8.

Terrestrial Comparison

Three- and four-sided pyramidal shapes are the most durable forms of construction, and their modeled triangular faces are not the product of random occurrence in a natural environment. Natural pyramidal formations tend to have conical shapes that are absent of equal or similarly sized faces. They are commonly thick, truncated mounds or resemble tall, slender cones that have a round base with a pointed apex.[21]

We are all familiar with the enormous four-sided pyramids produced by the ancient Sumerians, Egyptians, and the Maya; however, the architectural construction of a three-sided pyramid is extremely rare. With the advent of Google Earth, obscured pyramidal formations that were once unknown

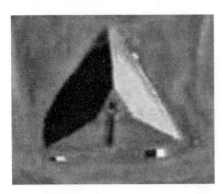

Fig. 1.9. Three-sided pyramidal formation (berm). Nevada National Security Site. Map data © 2007 Google.

to the general public have recently been revealed, most notably in the Nevada desert.

In 1951 approximately 1,360 square miles of land, located 65 miles north of Las Vegas in the remote desert, was withdrawn from public use by the United States Atomic Energy Commission. This vast plot of land, referred to as the Nevada Test Site or the Nevada National Security Site, became widely known as the home of Area 51.[22]

Within this highly restricted area is a three-sided pyramidal formation that is part of the Big Explosives Experimental Facility (Fig. 1.9). This triangular formation is a modern construction, referred to here as a berm.[23] The berm has three triangular faces that are constant in width along each side, and it has a flat, triangular peak. Like the Great Pyramid in Egypt, its cap stone is missing, leaving a flat, triangular platform at the top. I believe it is fair to say that this triangular formation looks a lot like the three-sided pyramid that Sagan saw in the original Mariner 9 image.

The Kite Pyramid

Just below the Three-Sided Pyramid is another highly symmetrical pyramidal formation; in the early spring of 2010 the MRO HiRISE camera captured a great image of it during the early evening with a resolution of 5.5 meters per pixel. The image shows an elongated, four-sided pyramid with sharp angles, labeled C in Fig. 1.4. The kite-shaped formation has a medial spine that runs through the entire structure, like a cross, creating a set of four almost symmetrical faces that are equal in size and shape (Fig. 1.10). The kite-shaped formation measures 4.35 miles in length from its westernmost point to its eastern side and is 4.20 miles wide from its westernmost point to its eastern side.[24]

Fig. 1.10. Kite Pyramid (labeled C in Fig. 1.4). Detail MRO CTX B20_017574_1965_XN_16N198W (2010).

Fig. 1.11. Kite Pyramid (labeled C in Fig. 1.4). Detail MRO HiRISE CTX J03_046136_1965_XN_16N198W (2016).

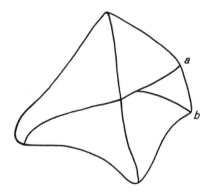

Fig. 1.12. Kite Pyramid. Analytical drawing with notations by the author.

A second image of the Kite Pyramid was acquired by the MRO HiRISE camera in the late spring of 2016 (Fig. 1.11). The picture was taken during the morning with a resolution of 5.5 meters per pixel. An analytical drawing highlighting the supportive spines of the Kite Pyramid is offered in Fig. 1.12. The sides of the formation have an undulating vertical ribbing feature that continues around the formation within each of its faces. Notice the central spinal ridge line deviates off its medial track (labeled *a* in Fig. 1.12) and curves down toward the south, creating a curving arch (labeled *b* in Fig. 1.12).

Terrestrial Comparison

In 2012, one of a new breed of satellite archaeologists working at the University of Alabama, Sarah Parcak, discovered an ancient polygonal mound in the Tunisia region of Northern Africa. It was discovered while she was mapping the paths of ancient Roman roads in the area. The polygonal mound, which dates back over two thousand years, was altered and transformed into a fortress by the Romans around 146 BCE[25] (Fig. 1.13). The contours of the mound have an elongated, polygonal shape that looks remarkably similar to the overall footprint of the Kite Pyramid observed on Mars (Fig. 1.11).

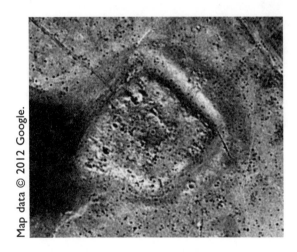

Map data © 2012 Google.

Fig. 1.13. Mound Fortress. Tunisia, Northern Africa.

The Star Shield

Moving a little further down, below the Kite Pyramid is another polygonal formation labeled B in Fig. 1.4. The MRO HiRISE camera acquired a great shot of this formation in the summer of 2010. The image was taken in the midmorning hours with a resolution of 5.5 meters per pixel.

The image shows this massive polygonal formation to be highly symmetrical with five angular points that resemble a star-shaped shield (Fig. 1.14). The soft blanket of morning light captures the sharp, cubic angles that appear to have been cut out of this massive mound, like butter. It seems to have been created with the precision of a sculptor or master architect.

Notice the sharp, triangular peak of the star-shaped structure and the supportive spines that run through its extended points, giving it an overall star shape. The top of the structure appears flat and has a kite shape, while its southeastern side has a triangular shape. Following its outer dimensions, the

Fig. 1.14. Star Shield (labeled B in Fig. 1.4). Detail MRO HiRISE CTX G01_018708_1959_XN_15N198W (2010).

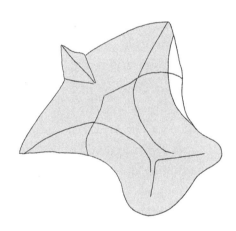

Fig. 1.15 Star Shield (labeled B in Fig. 1.4).
Left: Detail MRO HiRISE CTX F16_041969_1960_XN_16N198W (2015).
Right: Analytical drawing and gray wash by the author.

structure is almost square, and despite the partial collapse on its northeastern side, it is almost perfectly symmetrical. It measures 6.32 miles in width from its southwestern point to its adjacent northeastern point and 6.30 miles from the tip of its apex to its lower southeastern side.[26]

A second image of the Star Shield was acquired by the MRO HiRISE camera in the summer of 2015 (Fig. 1.15). The CTX image was taken in the early evening with a resolution of 5.5 meters per pixel, confirming the overall star shape of the formation and highlighting its sharp angles. An analytical drawing is provided in Fig. 1.15.

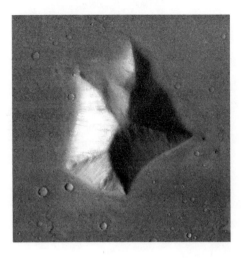

Fig. 1.16 Star Shield Jr. (labeled D in Fig. 1.4). Detail MRO HiRISE CTX P03_002318_1961_ XN_16N198W (2007).

Star Shield Jr.

Looking at the sparse plane just above the Kite Pyramid there is another polygonal, star-shaped formation (Fig. 1.16), labeled D in Fig. 1.4. Like its larger companion the Star Shield, labeled B in Fig. 1.4, this smaller version is also highly symmetrical with sharp, internal spines with cubic angles. The formation can be seen in the same MRO HiRISE CTX image that included the Three-Sided Pyramid in the winter of 2007. The bright early morning light that is coming in from the western side accentuates the light and dark areas of its interior angles. The formation has five points, and the northern point is off

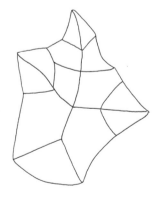

Fig. 1.17. Star Shield Jr. (labeled D in Fig. 1.4).
Left: Detail MRO HiRISE CTX D06_029600_1968_XN_16N198W (2012).
Right: Analytical drawing by the author.

set and curves to the west. Its peak is flat and has a small, square shape from which its supportive spines radiate.

A second MRO HiRISE CTX image of the Star Shield Jr. formation was obtained in the winter of 2012 (Fig. 1.17). The new image was taken in the late evening with a resolution of 5.6 pixels per meter. The overall size of the Star Shield Jr. formation measures approximately 4.10 miles from its northernmost point to its lower side and is 3.47 miles wide.[27]

Although taken five years later and at a different time of day, its lighting is very similar to the earlier MRO HiRISE CTX image taken in 2007. An analytical drawing of Star Shield Jr. is provided in Fig. 1.17.

Terrestrial Comparison

Various types of star-shaped, polygonal structures were developed as fortifications in the late fifteenth and early sixteenth centuries throughout Europe. A magnificent example exists in Uruguay, within the Department of Rocha, not far from the Atlantic Ocean. Built on a granite mound its construction was begun by the Portuguese in 1762 and eventually completed by the Spanish in 1775.[28] The fortress has a sharp, pentagonal form with five angled bastions that project out like arrows.

When the pentagonal templet of Fort Santa Tereza is compared to the pair of star-shaped structures found within the Sagan Complex on Mars, their common polygonal star design is remarkably similar.

Fig. 1.18. Fort Santa Tereza, circa 1776.

Map data © 2018 Google.

Dagger Mound

In the same early morning MRO HiRISE CTX image that captured the Star Shield Jr., there was included an elongated formation with a curved point that resembles a dagger (labeled F in Fig. 1.4). The formation has a rectangular shape, like a linear berm, however it curves off to the right and tapers down to a sharp point (labeled F in Fig. 1.4). The overall shape of the formation is highly geometric and symmetrical in design (Fig. 1.19). Notice its medial ridge, which extends across its entire length and runs right through its tapered point. It also has lateral supports that extend from its medial ridge, giving the formation its cubic shape. The formation is 3.67 miles long and 1.30 miles wide.[29]

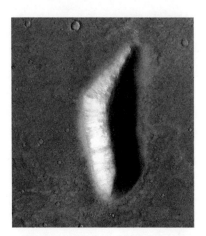

Fig. 1.19 Dagger Mound (labeled F in Fig. 1.4). Detail MRO HiRISE CTX P03_002318_1961_XN_16N198W (2007).

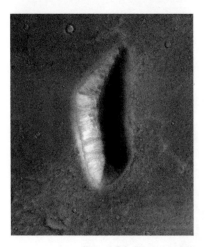
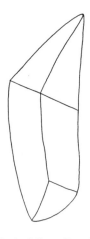

Fig. 1.20 Dagger Mound (labeled F in Fig. 1.4).
Left: Detail MRO HiRISE CTX P13_006142_1964_XN_16N198W (2007).
Right: Analytical drawing by the author.

A second image of the Dagger Mound is offered in another MRO HiRISE CTX image that was taken only ten months later (Fig. 1.20). This second image was taken in the early winter of 2007 during the morning hours, with a resolution of 5.5 pixels per meter. The lighting and resolution of the two available MRO HiRISE CTX images are almost identical; however, the second image provided a little more definition to the ridge lines separating its five faces. An analytical drawing is provided in Fig. 1.20.

Terrestrial Comparison

The Dagger Mound highly resembles the linear earthen mounds that have a rectangular or trapezoidal shape, known as a barrow. It is believed that these mounds were used as burial tombs in the early Neolithic period across Western Europe and England.

A similar large dagger-shaped barrow was first recorded by antiquarian William Stukeley in his eighteenth-century book, *Abury, a Temple of the British Druids, with Some Others, Described.*[30] (Fig. 1.21). The formation, known as Beckhampton Long Barrow, is located in Wiltshire, England, and is thought to have been constructed around 3200 BCE.[31] The size of the original formation is estimated to have been over 270 feet in length and 19 feet high.[32] Considering its curving dagger shape, it is interesting to note that the only artifact retrieved from within the mound was a bronze dagger that echoed the mound's shape.[33]

Fig. 1.21. Beckhampton Long Barrow. Wiltshire, England (circa 3200 BCE). Drawing by the author.

The Wiltshire site was surveyed and excavated again about a hundred years later by an English psychiatrist and part-time archaeologist John Thurnam in the mid-1800s. During his excavation and evaluation of the barrow, Thurnam reported that portions of the original dagger-shaped mound were partially destroyed and much of the western end was plowed away.[34]

Another earthen mound with a curved dagger shape was produced on the other side of the world, in the midwestern section of North America. Located in Madison, Wisconsin, near Lake Mendota, a large linear mound once existed that resembles the design of the Dagger Mound observed on Mars. The dagger-shaped mound was part of a late Woodland Period site known as the Hudson Park Mound Group that dates to 500 CE. The site originally included over two dozen mounds of linear and circular shape and had various effigy mounds depicting birds and unidentified animals.[35] The site was surveyed in 1888 by T. H. Lewis, and unfortunately all but one of the mounds were destroyed by residential development in the early 1900s. The only surviving mound at the site is one described as resembling a panther.[36] Without Lewis's original site map we would not have any evidence of this linear dagger-shaped mound (Fig. 1.22).

Geomorphology

There are no known geomorphological processes or mechanisms that could be responsible for producing these pyramidal and polygonal formations observed in the Sagan Complex (Fig. 1.4). After examining the area geologist Michael Dale maintains that although there is evidence that liquid water once flowed

Fig. 1.22. Dagger Mound. Detail of Hudson Park Mound Group, Wisconsin (circa 500 CE). Drawing by the author.

on Mars, which produced a variety of teardrop-shaped islands on the planet, there is no known fluvial processes that can produce these sharp-edged, polygonal formations that are symmetrical. The abrasive action of wind-borne particulates can also have a major effect on altering the geomorphology of a mound or knob by creating sharp angles. The most common result of this action is a yardang, which appears as a wedge-shaped hill that has sharp edges that align with prevailing winds.[37]

As an example, if we compare the common polygonal star shape of the Star Shield formation and its smaller companion, Star Shield Jr., although they are almost identical in their shape the Star Shield formation is set within a northwestern orientation, while Star Shield Jr. points in a northeastern direction. It is improbable that two almost identical star-shaped formations have been cratered in opposite directions. For wind to be responsible for creating this pair of star-shaped formations, the wind would have to shift to an opposing direction. However, each time the wind shifted to a new direction, it would erase the sharp edges created by the previous wind direction. As a result, these polygonal formations and their surrounding neighbors would all soon transform into an uninteresting group of circular mounds and buttes.[38]

Alignments

In examining the entire Sagan Complex, it became quite apparent that there were no clear linear alignments shared between the major formations that sit above the Sagan Pyramid. All the formations appear to be oriented in multiple directions (Fig. 1.23 on page 50). It was at this point that I took a closer look at the curvature of the northeastern star point of the Star Shield Jr. formation (Fig. 1.24 on page 50). Why was it bent toward the west? Was it intentional? Was it some kind of directional marker?

When the curvature of the northern star point of the Star Shield Jr. formation is followed along its westward trajectory, it aligns with the curved point of the Dagger Mound and rubs the outer rim of a small crater (labeled *a* in Fig. 1.25). The arch circles around and touches the northern tip of the Kite Pyramid and continues right up to where it began along the edge of the bent star point of the Star Shield Jr. formation. The arching alignments shared between these three structures fall within a circular template that is approximately fifteen miles in diameter.

Looking again at the Kite Pyramid, it also has an odd curvature to its medial ridge line (Fig. 1.26 on page 51). When a circular templet of the same arching

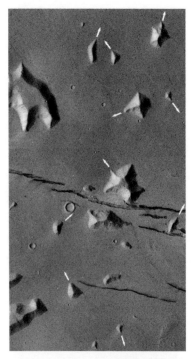

Fig. 1.23 The Sagan Complex. Detail ESA H5208_0000_ND3 (2008). Notated with directional arrows by the author.

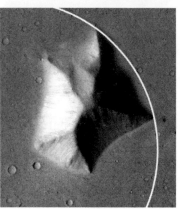

Fig. 1.24. Star Shield Jr. Detail MRO HiRISE CTX P03_002318_1961_XN_16N198W (2007). Arching circle by the author.

circle that connected the first set of formations is overlaid with the curvature of the Kite Pyramids medial ridge line, additional alignment points are revealed.

Following the trajectory of the arching circle it passes over the outer rim of a small crater labeled *a* in Fig. 1.27 and aligns with the slight curvature of apex point of the Star Shield formation. The arch continues down through the main body and out its southeastern point. The arch then passes the outer rims of two small craters labeled *b* and *c* in Fig. 1.27 and completes its path at the southeastern point of the Kite Pyramid. It then follows along the curvature of its medial ridge line and ends at its western point, exactly where it began.

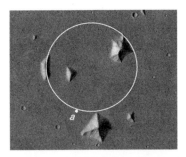

Fig. 1.25. Star Shield Jr. alignments. Detail ESA H5208_0000_ND3 (2008). Circle and a notation by the author.

Fig. 1.26. Kite Pyramid. Detail MRO HiRISE CTX J03_046136_1965_XN_16N198W (2016). Arching circle by the author.

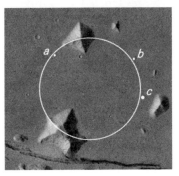

Fig. 1.27. Kite Pyramid alignments. Detail ESA H5208_0000_ND3 (2008). Circle and notated a, b, and c by the author.

Moving down to the original Sagan Pyramid, when the same circular template is placed over the northwestern and eastern points of the Sagan Pyramid (labeled a and b in Fig. 1.28) a third set of arching alignments is revealed.

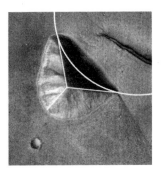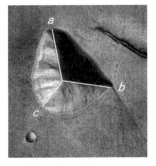

Fig. 1.28. Sagan Pyramid. Detail MRO HiRISE CTX P11_005219_1961_XN_16N198W (2007).

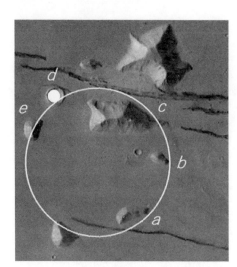

Fig. 1.29 Sagan Pyramid alignments. Detail ESA H5208_0000_ND3 (2008). Circle by author.

Following the arching trajectory from the Sagan Pyramid, its perimeter aligns with the outer edge of three other structures labeled *a*, *b*, and *c*, and the rim of a large crater labeled *d*. It proceeds around and cuts through the center of a mound labeled *e* (Fig. 1.29).

So what are the odds that the same circular templet could be placed over three different sets of formations and each set of formations produce connective alignments that fall within a common radius with a fifteen-mile diameter? Fig. 1.30 shows the totemic stacking of the three circular segments of the Sagan Complex, labeled 1, 2, and 3.

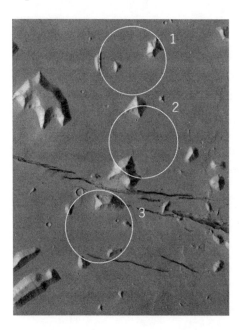

Fig. 1.30. Circular alignments within the Sagan Complex. Detail ESA H5208_0000_ND3 (2008). Circles and notations 1–3 added by the author.

Terrestrial Comparison

A vast assortment of earthen mounds and pyramidal structures can be found throughout the world from North and South America to Africa and Asia and across the entire European continent. The largest concentration of such structures was produced by the early Indigenous peoples of North America known as the Mound Builders. The early inhabitants of North America began constructing earthen mounds and pyramidal formations with geometric and animal forms over 3,400 years ago. Many of these formations were only a few feet high while others were as massive as small mountains. It is believed that these formations were built for religious, ceremonial, and burial purposes.[39] Many of the mounds created by this little-known culture are considered enormous feats of engineering. The remains of Monks Mound in the state of Illinois measures over 100 feet in height, 955 feet in length and 775 feet wide. Its circumference is comparable in size to the base line of the Great Pyramid of Giza.[40]

It is estimated that the number of earthworks produced throughout North America alone numbered in the hundreds of thousands.[41] Unfortunately, with years of neglect, almost all these earthen monuments have been either destroyed by natural erosion or by the rapid expansion of rural and urban development. Of all the mounds produced throughout the United States over ten thousand occupied the area found within the northeastern region of Iowa.[42] These mounds extend northward from Iowa into southeastern Minnesota and expand across the southern portion of Wisconsin into the boundaries of Illinois, Indiana, and Ohio. Others can be found along the shores of the Mississippi extending down into the southern states, while still more examples have been found in Alabama and Georgia.

In the mid 1800s a landmark study of these prehistoric mounds built throughout the Midwest was conducted by reputable scientists of the day. Most notably was the work conducted by a pair of American archaeologists, Ephraim G. Squier and Edwin H. Davis. Their work began in the forefront of the newly formed field of archaeology when they undertook one of the most ambitious archaeological endeavors ever attempted. The pair wanted to document as many of these earthworks before their remains fell victim to the ever-encroaching settlements that were plowing them under. From 1845 through 1847, the team surveyed nearly one hundred earthworks and hundreds of mounds, and in 1848 a collection of their field work was published in the *Smithsonian Institution's first volume of Contributions to Knowledge* series.[43]

The comprehensive fieldwork produced by Squier and Davis has left us with numerous drawings and highly detailed maps that provide contemporary archaeologists with a vast archive of earthen mounds. Their collective work not only allows researchers to reexamine these early settlements and cities but to locate numerous sites that no longer exist. Their exhaustive fieldwork also showed that the Mound Builders produced more than just simple circular and linear mounds. They produced massive earthworks in the form of everything from pictographic images of birds, deer, and bears to geometrical forms that take the precise shape of squares, rectangles, and ovals. The scale and complexity of some of these formations are truly engineering marvels.

Remarkably one of the sites surveyed by Squier and Davis has a set of seven mounds arranged around a central plaza, very similar to what we see on Mars (Fig. 1.31). The site known as the Jordan Mounds is in the Morehouse Parish sector of Louisiana. The mounds are located near the Arkansas River, and although they are thought to have been constructed during the 1500s they were found to be in an excellent state of preservation.[44]

Beginning at the center of the large mound labeled Mound A in Fig. 1.31, a circular arch can be followed from the lower, southwestern corner of the mound's top platform to its northern corner. The sight line of the arch hits the outer edge of Mound B as it circles around and cuts through the center of Mounds I, H, G, and F. From there it winds around and passes through the western corner of Mound E as it completes its path, ending at Mound A (Fig. 1.31).

Another example of a group of mounds set in a circular pattern was also observed by Squier and Davis. Located near the Black Warrior River in Alabama is a group of mounds produced by an early Mississippian culture that are arranged in a circular pattern. Known as the Moundville Site (Fig. 1.32), it is one of the largest sites in the United States. It covers four-and-a-half square miles and contains twenty-nine mounds that date to around 1000 CE.[45]

Beginning with the surrounding mounds that hover around the central mound labeled A, I'll start at the center of the mound labeled R at the top left of Fig. 1.32 on page 56. Following its circular trajectory, in a clockwise direction, it goes through the center of Mound C and proceeds along the eastern edge of Mound D. The arching circle continues through either the center or outer edges of the remaining mounds ending where it began with Mound R (Fig. 1.32).

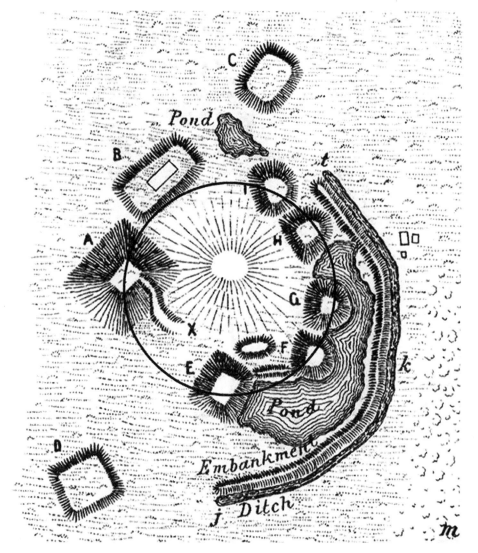

Detail of site map courtesy Squier and Davis, Plate XXXVIII, Figure 4 (1884).

Fig. 1.31. The Jordan Mounds, Louisiana, circa 1554.
Circle added by the author.

I'm confident that if Squier and Davis had the opportunity to examine and survey the pyramidal structures identified here within the Sagan Complex on Mars they would think they were looking at another site produced by the Indigenous Mound Builders of North America.

I'm also confident that if we follow the guidelines for artificiality provided by Carl Sagan in his book *Cosmos*, in which he states, "The first indication of intelligent life on Earth lies in the geometric regularity of its constructions,"[46] even Sagan would agree that these pyramidal structures on

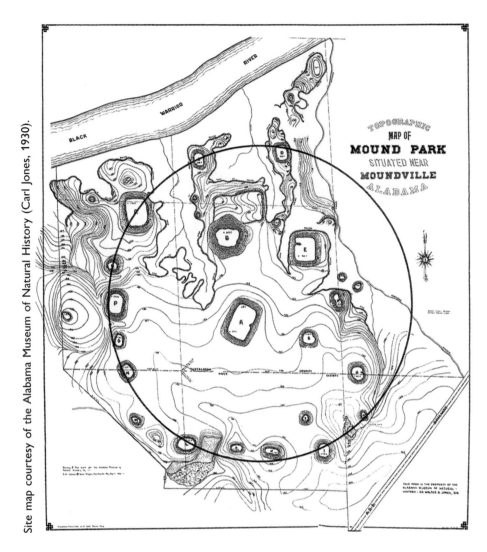

Fig. 1.32. Moundville. Hale/Tuscaloosa, Alabama, 1000 CE.
Circle added by the author.

Mars exhibit a direct correlation with the mound-building cultures explored by Squier and Davis.

Remarkably, this is not the only grouping of pyramidal formations on Mars that express such alignments. If Sagan had expanded his search further south to the outer edges of Elysium Planitia he would have been confronted with another Martian outpost of architectural wonders.

TWO

Darkness on the Edge of Forever I
Mean City: The Northern Territory

Mean Green

In April 2018 I was alerted to an interesting group of formations within the Nepenthes Mensae region of Mars by an independent researcher that goes by the moniker Mean Green.[1] He posted a portion of the area labeled 1, in Fig. 2.1, on his Facebook page. He also provided friends with a link to the original European Space Agency's Mars Express image H2081_0000_ND3. After I downloaded the full ESA image, I began exploring the area and noticed a variety of oddly shaped mounds and pyramidal structures seen throughout the image (Fig. 2.1 on page 58). I was so impressed by the concentration of so many anomalous formations within a single image, I started referring to the entire area as Mean City.

Nepenthes Mensae

Nepenthes Mensae is located near the equator in the eastern hemisphere of Mars. It has an expansive, rugged terrain that is located between the highlands of Terra Cimmeria and the low plains of Elysium Planitia. The surface area varies in height, and the surrounding terrain can exhibit a variety of textures, from smooth planes to rippling dunes and low ridges.[2] Recent observations in the area have revealed geomorphological features that could be related to a standing water sheet in the area, such as fluvial terraces, deltas, and shorelines.[3] Its coordinates lie between 14°N and 0.45°S and expands from 99°E to 134°E.[4] The formations

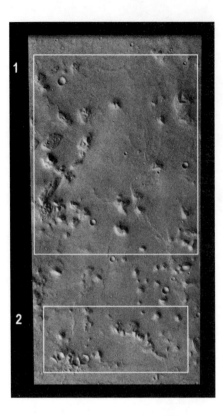

Fig. 2.1. The Mean City
Complex. Detail Mars
Viewer, ESA Mars Express,
H2081_0000_ND3 (2005).
1. The Northern Territory.
2. The Southern Territory.
Notations by the author.

of interest in this study are located at approximately 7°N and 124°E (Fig. 2.2). The word Nepenthes is derived from the Greek word, which means "without grief" or "sorrow." It is a reference to a passage found in Homer's Odyssey, in which a magical potion called "Nepenthes pharmakon" is given to Helen by an Egyptian queen to quell her sorrows with a cloak of forgetfulness.[5]

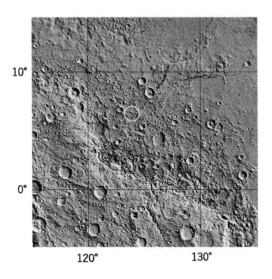

Fig. 2.2. Nepenthes Mensae
region of Mars (MOLA map).
Notated with approximate
location of the area of study
(circled).

Mean City: The Northern Territory

I will begin this portion of the study by utilizing the 2005 ESA, Mars Express image H2081_0000_ND3 (Fig. 2.1) that was introduced to me by Mean Green. The original image was acquired by the Mars Express camera in the summer during the late morning with a resolution of around 13 meters per pixel. The sun hits this collection of oddly shaped surface features from the west, providing a crisp light that defines their provocative shapes. In Fig. 2.3, I have identified some of the more interesting and geometrically shaped formations in the area and labeled them A through F.

Starburst

The first oddity to attract my attention was a highly irregular star-shaped formation located in the upper left-hand corner of the ESA Mars Express strip (Fig. 2.4 on page 60). The formation has a polygonal form that has five projectile points with a truncated top. The surface of the formation is mostly flat, with five radiating arms that stretch out like a giant starfish. There is a large mound positioned at its northernmost point and three smaller mounds of various sizes, located at its center. The formation projects so much energy in its shape and design that I have titled it Starburst.

In 2016 the Mars Reconnaissance Orbiter mission released a HiRISE CTX image that included the area around the Starburst formation (Fig. 2.5 on page 60). The image was taken in the late spring, during the early afternoon, with a resolution of 5.8 meters per pixel. Although taken eleven years later and

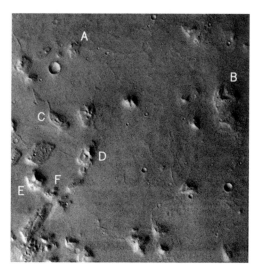

Fig. 2.3. The Mean City Complex: Northern Territory. Detail ESA H2081_0000_ND3 (2005). Notated A–F by the author.

Fig. 2.4. Starburst (labeled A in Fig. 2.3). Detail ESA H2081_0000_ND3 (2005).

Fig. 2.5. Starburst (labeled A in Fig. 2.3). Detail MRO HiRISE CTX J04_046177_1871_XN_07N235W (2016).

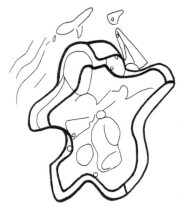

Fig. 2.6. Starburst. Analytical drawing by the author.

under slightly different lighting conditions this image supports its proposed star shape and confirms its contours and interior morphology. An analytical drawing is provided in Fig. 2.6, highlighting its contours.

Terrestrial Comparison

The polygonal, wedge-shaped Starburst formation on Mars highly resembles a star fort with triangular bastions at each corner. Star-shaped forts were commonly found in Europe, and their design was adopted during the colonization of America as well as the Civil War period. Many of these star-shaped fortifications included interior buildings and had raised platforms within their main structure allowing military fire over the main ramparts.[6]

After going through dozens of examples of star forts I came across a highly comparable fortification known as Fort Henry (Fig. 2.7). The military fortress was named after its designer, a Confederate senator of Tennessee, Gustavus Henry. The fort was designed as a five-sided, open-bastioned earthen structure that expands over ten acres. It was built in 1861 on the eastern bank of the Tennessee River to moderate traffic on the river.[7]

When Fort Henry is compared to the Starburst structure found on Mars their common polygonal star design is remarkably similar (Fig. 2.7). Notice the various sizes and shapes of the extending bastions of Fort Henry and its truncated star point at the top. It is this truncated section of Fort Henry that looks very similar to the blunted star point observed on the Starburst structure on Mars.

Unfortunately, throughout most of its long history Fort Henry was abandoned and decommissioned to a private owner and the site was never maintained or preserved. In 1944 after the Tennessee River was dammed by the

Fig. 2.7. Common star design.
Left: Fort Henry, Tennessee (1861).
Note its flat top point is oriented to the west.
Right: Starburst (labeled A in Fig. 2.3). Detail MRO HiRISE CTX
J04_046177_1871_XN_07N235W (2016).

Tennessee Valley Authority, creating the Kentucky Lake, the remains of this earthen fortification were submerged under water.[8] Perhaps its companion on Mars suffered a similar fate?

Open Delta

Looking to the eastern side of the ESA Mars Express strip, directly across from the Starburst structure in Fig. 2.3, I noticed a symmetrical U-shaped, triangular mound. The formation has a flat bottom, giving it a somewhat triangular, open delta design (Fig. 2.8). The formation has soft, rounded edges and corners with an open peak, giving it an overall impression of an isolated seaport or jetty in the shape of an open delta.

Here is a section of an image that was released in 2012 showing another view of the Open Delta formation. It was acquired in the spring, during the early afternoon, with a resolution of 5.4 meters per pixel.[9] This second image provides more detail to the cluster of elongated mounds seen within the upper left arm and along the central, horizontal band of the triangle imprint. The horizontal band has segmented folds that resemble a breakfast croissant, flanked by a pair of large craters. The craters appear to have caused a massive disturbance to the area below the horizontal band. Notice the ejecta debris flowing down in a southwestern direction creating a V-shaped blanket. The left arm of the Open Delta formation is highly elevated with thick, supportive embankments, while the right arm is almost level with its adjoining surface. The right arm is pitted, and its edges are partially covered in sand and debris. Notice the evenly spaced, linear grid marks on the surface that extend out toward the northeast like a plowed field. An analytical drawing of the Open Delta structure is provided for review in Fig. 2.10.

Fig. 2.8 Open Delta (labeled B in Fig. 2.3). Detail ESA H2081_0000_ND3 (2005).

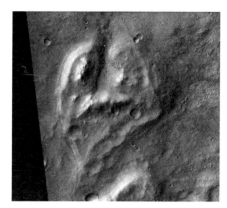

Fig. 2.9. Open Delta (labeled B in Fig. 2.3). Detail MRO HiRISE CTX G21_026542_1894_XN_09N235W (2012).

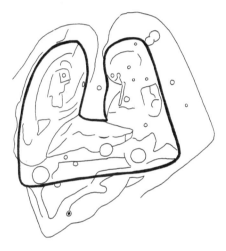

Fig. 2.10. Open Delta. Analytical drawing by the author.

Evidence has shown that there were water-related landforms in Nepenthes Mensae in the past,[10] and the Open Delta formation may have been a large landmass surrounded by water. The splayed, open area at the head of the formation would allow water to flow in from the surrounding waterway creating access to a port with double jetties. At some time in the past, its southern horizontal band was damaged by two targeted impact events.

Terrestrial Comparison

Searching through modern architectural designs of ports and inland structures I came across two examples that show that the U-shaped form of the Open Delta structure on Mars exhibits a feasible design. The first is a port proposed by the Vietnam Port Consultant Corporation, which is called the Cai Mep Ha Logistics Center[11] (Fig. 2.11 on page 64). Notice the U shape of the port and the extended docks of slightly different lengths.

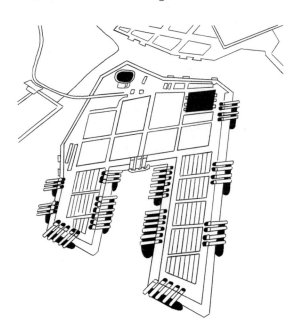

Fig. 2.11. Cai Mep
Ha Logistics Center.
Proposed designed
by Vietnam Port
Consultant Corporation.
Drawing by the author.

The second candidate for comparison is an open-ended triangular building designed by the Dutch architectural firm Jo Coenen, which is based in
Amsterdam[12] (Fig. 2.12). Notice his architectural plans show an asymmetrical
U-shape building with one arm slightly longer than the other. I must say that
both of these examples look very similar to the Open Delta formation on Mars.
The visionary concept shared between these three architectural structures is
quite remarkable.

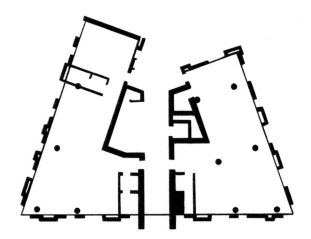

Fig. 2.12. Open Triangle building (architectural plans).
Design by Jo Coenen and Archisquare in Parma.
Drawing by the author.

Bulldog Bat Head

Returning to the western side of the ESA Mars Express image H2081_0000_ND3 shown in Fig. 2.3, there is a teardrop-shaped mound that I have labeled C. It sits just below the Starburst structure and is in direct alignment with the Open Delta structure on the eastern side. The mound has a flat, knifelike shape with two wavy linear mounds (Fig. 2.13). Its southeastern end appears puffy and has a curved, fish-hook shape. Most of its interior features are muted and obscured by dark shadows.

A second view of Structure C was acquired by the MRO HiRISE spacecraft that provides a much better view of its topographic features (Fig. 2.14). The image was taken during midmorning in the winter of 2010 with a resolution of 5.4 meters per pixels. The early morning light provides additional detail to the surface features, which resemble the head of a bulldog-faced bat. The new image highlights a wedge-shaped ear and decorative ear lobe. It also reveals a recessed eye socket that is embellished by a flaring brow, which extends up like a blooming flower along the ear. There is even evidence of a small, cherry-size nose and nostrils at the end of its muzzle. The lower hooked end of the formation takes on the form of gaping mouth. Notice the firm jaw and squared

Fig. 2.13 Structure C (labeled C in Fig. 2.3). Detail ESA H2081_0000_ND3 (2005).

Fig. 2.14. Bulldog Bat Head (labeled C in Fig. 2.3). Detail MRO HiRISE CTX B17_016270_1878_XN_07N236W (2010).

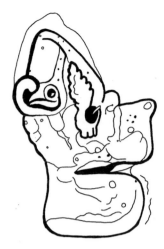

Fig. 2.15. Bulldog Bat Head.
Analytical drawing by the author.

muzzle. An analytical drawing of the Bulldog Bat Head is provided in Fig. 2.15 for review.

Terrestrial Comparison

To confirm my proposed identification of Structure C as that of a head of a bulldog bat, I searched through various images of bull-faced bats. I found the bulldog bat is part of the Noctilionidae family that consists of only two species, the greater and the lesser bulldog bat.[13] A profiled headshot of a greater bulldog bat matches the geoglyphic portrait of its companion on Mars (Fig. 2.16), which is possibly one of the ugliest species of bats I have ever seen.

Fig. 2.16. Greater bulldog bat, profile. Drawing by the author after photograph by Merlin Tuttle at Science Source Stock Photo.

The greater bulldog bat has a light brown to orange colored fur, and its body measures around five inches in length. These small bats have long legs with large feet and strong claws to grab prey. Their facial features include prominent eyes, large funnel-shaped ears, and a small nose. Their most identifiable attributes include wrinkled cheek pouches and full lips that are divided by a fold of skin giving them their bulldog appearance. Although these bats eat insects, their chief food source is fish.[14] These small bats use echolocation to locate fish swimming just below the surface of the water and then grab them with their long, sharp claws.[15]

Armored Fish

Toward the southeastern side of the Bulldog Bat Head geoglyph, in the ESA Mars Express image (labeled C in Fig. 2.3), there is a cluster of small, segmented mounds that take on the form of an aquatic creature. Taken together the contours of these conjoined mounds resemble a highly stylized armored fish (Fig. 2.17).

The undulating creature has a flat, curved tail fin and a segmented body beginning with a rounded abdomen that has an extended side fin. The thick, bulbous head appears higher than the main body, while its snout maintains a lower level with the main body. The eye area appears as a blank patch with radiating rays, while its snout is blunted like those seen on a killer whale. Notice the open mouth showing the remains of upper and lower teeth. An analytical drawing is provided in Fig. 2.18 on page 68 for review.

Over the years the MRO HiRISE spacecraft had the opportunity to retrieve two additional images of the Armored Fish but, unfortunately both images

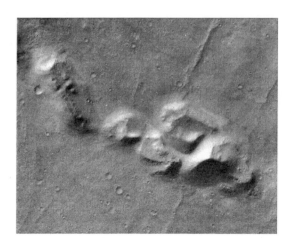

Fig. 2.17. Armored Fish (labeled D in Fig. 2.3). Detail ESA H2081_0000_ND3 (2005).

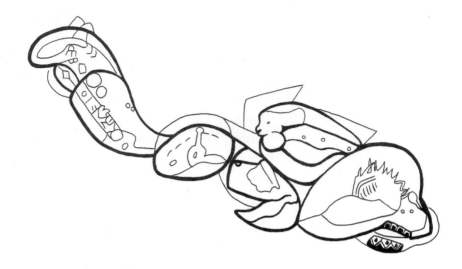

Fig. 2.18. Armored Fish. Analytical drawing by the author.

only capture partial views of the formation. The better of the two images is the one shown in Fig. 2.19, which captured the area surrounding its head. The image was taken in the spring, during the late evening, and was released in 2018 with a resolution of 5.4 pixels per meter. Despite being only a partial view, the best thing about this new image is that it was taken at a higher resolution than the earlier ESA Mars Express image, and it provides much more detail to the mouth and teeth.

In January 2020 the MRO HiRISE CTX camera acquired another image of the Armored Fish, this time capturing about 95 percent of the entire forma-

Fig. 2.19. Armored Fish (head detail). Detail MRO HiRISE CTX K05_055473_1871_ XN_07N236W (2018).

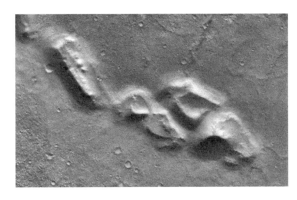

Fig. 2.20. Armored Fish (labeled D in Fig. 2.3). Detail MRO HiRISE CTX N02_063214_1882_XN_08N236W (2020).

tion, only missing a small portion of its tail (Fig. 2.20). Unlike the people controlling the ESA Mars Express spacecraft, it seems the MRO team has difficulty capturing the entire formation. The new image was taken in the winter, during the early morning, with a resolution of 5.4 pixels per meter.

Terrestrial Comparison

The highly stylized, segmented appearance of the Armored Fish shares a lot of its design elements seen in representations of killer whales often produced by Northwest Coast Indians of North America (Fig. 2.21). Notice the segmented body with embedded glyphs that form its side fin, dorsal fin, and curved tail fin. The composite creature has a bulbous head with a large eye patch and a slightly parted mouth showing its sharp teeth.

Fig. 2.21. Killer Whale. Northwest Coast Indian (Haida). Drawing by the author after Clarence Mills.

The killer whale is one of the most prominent subjects to be found in the art produced by the Northwest Coast Indians. The killer whale is regarded as a great hunter and a guardian as well as the ruler of the sea because of its sheer size and power.[16]

It is believed that when a fisherman's boat is capsized by a killer whale, as the fisherman sinks to the watery realm below, he is transformed into a killer whale. When the coastal people see a killer whale swimming near the shore, they believe it to be one of their lost family members trying to communicate with them.[17]

After finding this large geoglyphic formation of the Armored Fish and the profiled head of a seafaring Bulldog Bat resting on the surface above it, and seeing the starfish-shaped Starburst formation at the top of the group, I wondered if there was some kind of aquatic theme developing here.

Polygonal Step Pyramid and Fancy Dove

Continuing my exploration down the western side of the ESA strip in Fig. 2.3, the next oddity to be examined is a multitiered, five-sided polygonal step pyramid. It sits right next to an avian-shaped formation that looks like a fancy-tailed dove (Fig. 2.22). The interior morphology of the polygonal pyramid has a mix of smooth, lumpy forms that are dispersed around sharp-edged blocks that create a stepped appearance. The western and southern side of the formation is highly illuminated, while the northeastern side is quite dark.

Another view of this odd pair of formations was included in MRO HiRISE CTX image B17_016270_1878_XN_07N236W shown in Fig. 2.23. The illumination of this second image is more evenly dispersed and exposes a lot more structural detail. It shows stark sunlight hitting the multitiered steps that spiral

Fig. 2.22. Polygonal Step Pyramid and Fancy Dove (labeled E and F in Fig. 2.3). Detail ESA H2081_0000_ND3 (2005).

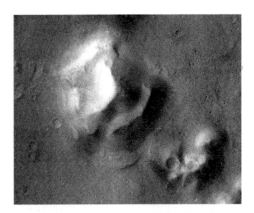

Fig. 2.23. Polygonal Step Pyramid and Fancy Dove (labeled E and F in Fig. 2.3).
Detail MRO HiRISE CTX B17_016270_1878_XN_07N236W (2010).

Fig. 2.24. Polygonal Step Pyramid (labeled E in Fig. 2.3). Detail MRO HiRISE CTX B17_016270_1878_XN_07N236W. Notice its symmetry. Outlined by the author.

around this polygonal formation while casting soft shadows across its interior form, giving it an undulating, chaotic topography. Notice the bulbous area in the center and the lower southern side that swings around with a flat rectangular ramp on its eastern side.

Along the outer contours of the Polygonal Step Pyramid the first thing that becomes obvious is that its outer perimeter is highly symmetrical and has a distinct five-sided, pentagonal star shape (Fig. 2.24). Masonic traditions see the five-pointed pentagonal shape as the Blazing Star. It is seen as a symbol of perfection and was associated with the followers of the Greek philosopher Pythagoras. They perceive the pentagram as the key to higher knowledge, which opened the door to secrets.[18]

Looking now to the lower right-hand side of the Polygonal Step Pyramid toward the east, I will examine the physical features of the Fancy Dove (Fig. 2.25 on page 72). In this second image, the midmorning light adds form to the dove's round head and puffed-up, feathered body. The eye and beak are readily seen in the flood of sunlight, while the flowing tail form appears

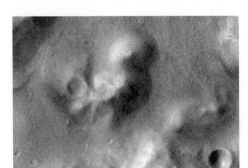

Fig. 2.25. Fancy Dove. Detail MRO HiRISE CTX B17_016270_1878_ XN_07N236W (2010).

Fig. 2.26. Fancy Dove. Analytical drawing by the author.

darker. This may be due to the tail's lower elevation, or the morning light is being blocked by the large pentagonal form sitting behind it.

The features of the Fancy Dove are exquisitely depicted. It has a round head with a small eye and triangular beak. The head is tucked down into a fully puffed-up chest that leads down to a footless body. The body extends into a large, fanned-out fancy tail that includes the circular remains of a previous crater hit. The dove's feet are difficult to confirm and may be obscured by sand and debris at its base. An analytical drawing of the Fancy Dove is provided in Fig. 2.26 for comparison.

Terrestrial Comparison

Starting with the Polygonal Step Pyramid, when its complex, multitiered design is compared to the vast variety of pyramidal styles produced around the world, a Mesoamerican example stands out. Located in western Belize, near the Guatemala border, is the little-known site of Xunantunich, which sits on a

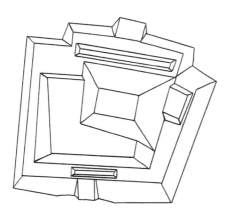

Fig. 2.27. Common polygonal step design.
Left: Polygonal step pyramid. Xunantunich, Belize. Drawing by the author.
Right: Polygonal Step Pyramid, Mars. Detail MRO HiRISE CTX
B17_016270_1878_XN_07N236W.

ridge high above the Mopan River. The site occupies about one square mile of land and consists of multiple plazas with large temples and palaces.[19] One structure has an interlocking stepped construction that resembles the Polygonal Step Pyramid observed on Mars.

Here is a comparison of a Mayan pyramid at Xunantunich with the Polygonal Step Pyramid on Mars (Fig. 2.27). Notice the tiered platforms and the blocky, overlapping levels that spiral around each pyramid.

Shifting over to the Fancy Dove geoglyph, I have identified this avian creature as a dove, because there is no scientific difference between doves and pigeons other than size and color variation. In the *Dictionary of Birds*, its author Alfred Newton says, "no sharp distinction can be drawn between pigeons and doves, and in general literature the two words are used almost indifferently."[20] Considering the overall aquatic theme emerging here, I thought the presentation of a dove could not be more appropriate.

Doves and pigeons are part of the Columbidae family of over three hundred species that are descended from the rock dove and are commonly found in just about every part of the world. The basic physical morphology of the common dove and pigeon are indistinguishable. Both birds have stout bodies with short necks. They have a smooth, round head and a short, slender bill that is highlighted by a fleshy cere. Their tails are normally short, while the more flamboyant varieties have developed a fancier, much longer and fuller plumage.[21]

Fig. 2.28. Fancy doves.
Top: Blue Grizzled Fantail. Drawing by the author after a photo by Lisa Adriana.
Bottom: Seldschuk Fantail. Drawing by the author after a photo by Jim Gifford.

The Fancy Dove geoglyph on Mars appears to be a fine example of one of the more flamboyant, fancy varieties of doves and pigeons that are found here on Earth (Fig. 2.28). The first breed to come to mind is the Blue Grizzled Fantail, which originated in India. It is a beautiful variety with a full, plump breast and a distinct fan-shaped tail with feathered legs and feet. The head is round with fluffy plumage, and it has a small, dark beak.[22]

A second variety for review is the Seldschuk pigeon, which comes from Anatolia, Turkey. It is a highly ornamental breed that has a full, plump breast and a distinct fan-shaped tail, while its legs and feet are normally bare. Its head is round and smooth with a medium-size beak and curved forehead.[23]

Taking another look at the juxtaposition of the Fancy Dove and the

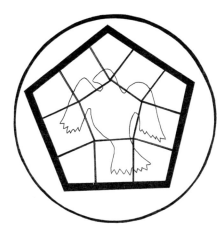

Fig. 2.29. Pentagon Caged Dove,
Button (circa 1967).
Drawing by the author.

Polygonal Step Pyramid, why are we seeing a dove sitting next to a pentagonal pyramid? Well, traditionally the dove is seen as a symbol of peace, and the pentagonal star shape has been seen as a symbol of power and war. Just as we saw with the defensive design of the Starburst formation at the top of this complex, the pentagon was also used to create a star fort. The home of the United States Department of Defense, which is better known as the U.S. Pentagon, has the same shape. Is it possible that these two formations on Mars suggest a similar war and peace scenario?

In 1967 a protest button was created by the National Mobilization Committee to End the War in Vietnam. The button shows a white dove trapped within a black pentagonal cage[24] (Fig. 2.29). The image suggests that peace is hopelessly suppressed by the endless power of the military industrial complex. The scenario we see on Mars expresses a more positive pairing, where an uncaged dove is free to fly away from war. Or perhaps the Polygonal Step Pyramid was meant to represent a sanctuary, something like an ark with the Fancy Dove preparing to take flight in search of the safety of fertile land.

A possible answer may lie to the south, just along the illuminated edge of darkness.

THREE

Darkness on the Edge of Forever II
Mean City: The Southern Territory

A Tightly Knit Train of Formations

RETURNING TO THE EUROPEAN Space Agency's Mars Express image H2081_0000_ND3, which captured the Mean City complex that was discussed in the previous chapter, I will now examine the area located slightly to the south, just below the Northern Territory (labeled 1 in Fig. 3.1). Moving down and across a sparse section of the landscape, there is a tightly knit succession of mounds that appear to form a train or cluster of pyramidal and geometrically shaped formations. The entire train of formations occupies an elevated platform that undulates across the terrain in a northwestern to southeastern direction (labeled 2 in Fig. 3.1). It is extremely interesting to find such a diverse set of geometrically shaped formations sitting side by side in the same area. It is also hard to imagine that such a diverse set of formations were formed by natural processes, because as we know, nature tends to make random, asymmetrical forms that are similar in size and shape.[1]

I will begin this portion of my study by taking a closer look at the tightly knit train of geometrically shaped formations set within the Southern Territory of the Mean City Complex. I have titled this group of formations in the Southern Territory the Conjoined Temple Complex. Notice this entire group of highly geometric formations all appear to be resting on slightly elevated sets of platforms that continue throughout the complex. I have divided this conjoined train of formations into four individual tracts that are labeled 1–4 in Fig. 3.2.

76

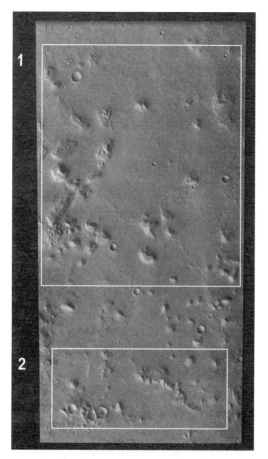

Fig. 3.1. The Mean City Complex.
Detail ESA Mars Express,
H2081_0000_ND3 (2005).
1. The Northern Territory.
2. The Southern Territory.

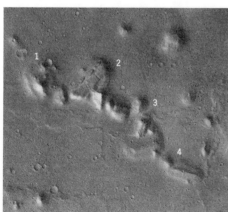

Fig. 3.2. Conjoined Temple
Complex (labeled 2 in Fig. 3.1).
Detail of ESA Mars Express
H2081_0000_ND3 (2005).
Notated 1–4 by the author.

Another image of the Conjoined Temple Complex was acquired by the Mars Reconnaissance Orbiter spacecraft in 2016. This second MRO HiRISE CTX image was taken in the summer during the early afternoon with a resolution of 5.8 meters per pixel. (Fig. 3.3 on page 78). It confirms the shape and

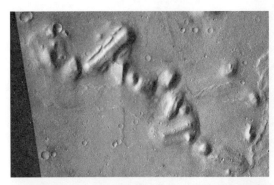

Fig. 3.3. The Conjoined Temple Complex (labeled A in Fig. 3.2). Detail MRO HiRISE CTX J04_046177_1871_XN_07N235W (2016).

Fig. 3.4. Proposed geometry within the Conjoined Temple Complex. Detail MRO HiRISE CTX J04_046177_1871_XN_07N235W (2016). Contours outlined by the author.

contours of all the formations observed in the earlier ESA Mars Express image, and it provides a better tonality and clearer resolution.

Here is an outlined version of the MRO HiRISE CTX image showing the proposed geometry of each structure. Notice the outer contours of this tightly knit cluster of geometrically shaped formations create an assortment of polygonal shapes, such as circles, ovals, octagons, triangles, and rectangles (Fig. 3.4). The entire complex measures a little over two miles from the tip of its northwestern side to its southeastern edge.[2]

Tract 1

Beginning with the northwestern section of the Conjoined Temple Complex I will focus on a group of formations that I have labeled Tract 1 in Fig. 3.2. Another image of the area was acquired by the MRO HiRISE CTX spacecraft in 2019 that includes three of the four sections of this traversing complex, showing much more detail. This third MRO HiRISE CTX image (Fig. 3.5) was acquired during the winter, in the midmorning, with an exceptional resolution of 5.4 meters per pixel.

In an effort to provide the structural clarification of each of the geometri-

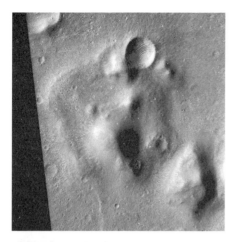

Fig. 3.5. Tract 1 (labeled 1 in Fig. 3.2). Detail MRO HiRISE CTX N01_062858_1871_XN_07N236W (2019).

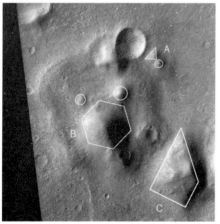

Fig. 3.6. Proposed geometry, Tract 1 (labeled 1 in Fig. 3.2). Detail MRO HiRISE CTX N01_062858_1871_XN_07N236W (2019). Contours outlined and notated A, B, and C by the author.

cal formations that occupy Tract 1, I have outlined their contours and labeled them A through C in Fig. 3.6. Notice the small triangular mound at the top of the image with an adjoined circular mound (labeled A). The mound sits below a large-impact crater with a sharp rim. The triangular and circular mounds are surrounded by a heaped accumulation of smooth sand and debris. At the center of the group is a large hexagonal mound (labeled B) that has two circular mounds above it. One mound is positioned to its left and one on its right, like bookends. Just beyond the eastern side of this hexagonal mound is a pair of small craters with erratic rims and spattered aprons. The eastern impact crater appears to have hit the southern corner of the hexagonal mound, causing only slight damage. This entire group of formations sits within an ejecta blanket of the large-impact crater seen at the upper right-hand corner of the platform.

Resting directly below this group, beyond the ejecta blanket, is a large wedge-shaped mound. The formation has four sides of various lengths, and it points up in a northern direction (labeled C in Fig. 3.6). It appears that this wedge-shaped mound and all the other geometrically shaped formations above it were constructed after the flow of the ejecta blank covered the area.

Tract 2

Moving on to the eastern side of Tract 1 is the second quarter of the Conjoined Temple Complex, which I have titled Tract 2 in Fig. 3.2. Another view of this section of the complex was acquired within the same image that captured Track 1 (Fig. 3.7). The image confirms all the geometric formations observed in the ESA Mars Express image.

I have labeled the geometrical-shaped formations in this section A through D, and their contours have also been outlined for structural clarification in Fig. 3.8. Notice the long, rectangular, bar-shaped mound labeled A in Fig. 3.8 is oriented in a northeast and southwest direction. Its interior is divided into two sections, with the lower section being twice as large as the northern section. It sits on its platform like a raised embankment with each of its ends slashed diagonally like a knife blade. There is also a set of fragmented, linear formations running along its northwestern side that are half buried and partially degraded by erosion.

Directly below the rectangular, bar-shaped mound is an elongated, triangular pyramid (labeled B in Fig. 3.8) that points directly south. To the eastern side of the triangular pyramid is an oval-shaped mound (labeled C in Fig. 3.8) that sits next to a larger, rectangular boxlike mound (labeled D in Fig. 3.8). The rectangular mound is about twice the size of the oval-shaped mound, and its topography has a scalloped ripple pattern that forms four riblike waves.

I find it very interesting that this section of the Conjoined Temple Complex begins with a slashing rectangular, bar-shaped mound (labeled A in Fig. 3.8) and proceeds along with three geometrically opposing forms. It is

Fig. 3.7. Tract 2 (labeled 2 in Fig. 3.2). Detail MRO HiRISE CTX N01_062858_1871_ XN_07N236W (2019).

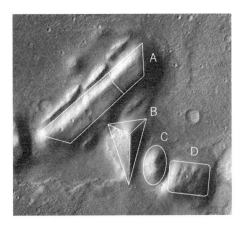

Fig. 3.8. Proposed geometry, Tract 2. Detail MRO HiRISE CTX N01_062858_1871_ XN_07N236W (2019). Contours outlined and notated A–D by the author.

puzzling as to how such a diverse set of formations occupy the same space. The group starts with a sharp, triangular pyramid that has two sides of equal length (labeled B in Fig. 3.8). It sits next to a soft, oval-shaped mound that is highly reflective (labeled C in Fig. 3.8). The group ends with a thick, rectangular box-like mound (labeled D in Fig. 3.8). The top of the boxlike mound is ribbed with three shallow ridge marks, and it has a lot of sand and dirt accumulated around its base.

Tract 3

Examining the third quarter of the Conjoined Temple Complex there is another set of geometrically shaped formations that I have titled Tract 3 in Fig. 3.2. To provide structural clarification I have outlined the contours of each of the formations and labeled them A–E in Fig. 3.9 on page 82.

This section begins with an oddly shaped mound that has soft contours, which appear to combine three conjoined circular mounds that are conflated with a large triangular form (labeled A in Fig. 3.10 on page 82). Notice the triangular form has blunted corners, and its upper side fans out, conforming to the outer contours of the upper, circular mound. The triangular form is thick and consists of light material, which causes it to be highly reflective. The base of the conflated formation sits on a flat, sand-scattered apron that arches around the structure and forms a hashtag, linear pattern.

Directly below the Triangular Mound with Conjoined Circles there is a uniquely shaped formation that is highly irregular. It has a W-shaped beak at its base and is topped with a tripronged feature that gives it the appearance of a Crowned Kite (labeled B in Fig. 3.10). Just to the eastern side of the

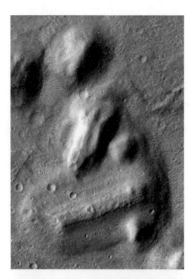

Fig. 3.9. Tract 3. Detail MRO HiRISE CTX N01_062858_1871_XN_07N236W (2019).

Fig. 3.10. Proposed geometry, Tract 3 (labeled 3 in Fig. 3.2). Detail MRO HiRISE CTX N01_062858_1871_XN_07N236W (2019). Contours outlined and notated A–E by the author.

Crowned Kite is a small, box-shaped mound (labeled C in Fig. 3.10) that is oriented in a southwest to northeastern direction. Directly below it is a long, rectangular, bar-shaped structure with triangular ends (labeled D in Fig. 3.10). It looks like a junior version of the Rectangular Bar-Shaped Mound, labeled A in Fig. 3.8. This smaller version is also highly symmetrical and sits on a slightly elevated platform that echoes its overall rectangular shape. Directly below this rectangular, bar-shaped mound is a small, square-shaped mound (labeled E in Fig. 3.10) that is aligned in a similar orientation. Notice the parallel, southwest to northeast alignments of the three formations, labeled C, D, and E in Fig. 3.10, which share the same platform.

One of the most interesting and highly unusual formations to occupy this set of formations observed in Tract 3 is the Crowned Kite. Utilizing the MRO HiRISE CTX image shown in Fig. 3.3 I took a much closer look at the formation's highly symmetrical design. The Crowned Kite has a sharp, W-shaped beak attached to a segmented, rectangular body that has a sharp, pointed tail fin that is flanked by a pair of extended arms that give it a W shape. Fig. 3.11 provides an outline of the structure's proposed contours.

Tract 4

Looking to the southeastern section of the Conjoined Temple Complex in the ESA image in Fig. 3.1 I will now examine the last section of this complex that I have titled Tract 4 in Fig. 3.2. The main formation appears as a large, tongue-shaped platform. It has a long, spinal ridgeline that extends across its center, while a large, circular mound sits on its northwestern side. There is also a small, oval mound located on the southeastern edge of its supportive platform.

Another view of Tract 4 was obtained in the winter of 2018 that provides additional clarity to the internal topography of this tongue-shaped platform (Fig. 3.12). This MRO HiRISE CTX image was acquired in the early evening with a resolution of 5.4 pixels per meter (Fig. 3.12).

Due to the lower sun angle of this image, it provides a better view of the formations that make up Tract 4. Fig. 3.13 provides an outlined version of the formations, highlighting their proposed geometry. The circular mound that occupies the top of the formation shows evidence of four, equally spaced, supportive spines set within a square form that sits on a circular apron (labeled A in Fig. 3.13). This image also shows more width to the spinal ridgeline that

Fig. 3.11. Proposed geometry, Crowned Kite (labeled B in Fig. 3.10). Detail MRO HiRISE CTX J04_046177_1871_XN_07N235W (2016). Outlined contours by the author.

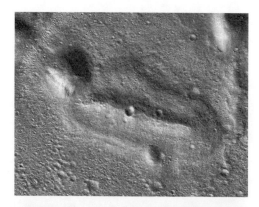

Fig. 3.12. Track 4 (labeled 4 in Fig. 3.2). Detail MRO HiRISE CTX K11_057768_1854_XN_05N235W (2018).

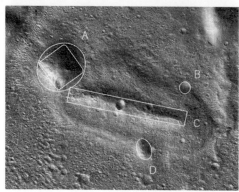

Fig. 3.13. Proposed geometry, Tract 4 (labeled 4 in Fig. 3.2). Detail MRO HiRISE CTX K11_057.768_1854_XN_05N235W (2018). Contours outlined by the author.

runs along the center of the platform, giving it a thicker rod shape (labeled C in Fig. 3.13). There is also a shallow impact crater located within the center of the spinal ridgeline. The eastern side of the tongue-shaped platform has a pair of circular forms along its northern and southern sides. The first is a shallow crater on its northeastern edge (labeled B in Fig. 3.14) and the other is a small, almond-shaped mound attached to its southeastern side (labeled D in Fig. 3.14).

Although the integrity of the structures set within the main platform may have been subject to an accumulation of mud and sediment over eons of time and then suffered the effects of wind and sand erosion, their basic, geometric form is still intact and visible.

Terrestrial Comparison

As I scan the entire group of geometrically shaped formations set within each tract of the Conjoined Temple Complex (Fig. 3.4), I am confident in saying that there are no terrestrial examples of conjoined platforms that are occupied

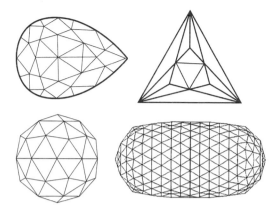

Fig. 3.14. Dome Arcology Model showing various geometric shapes. Graphic by the author

with geometrically shaped formations found anywhere in the natural landscape. They just do not exist. If a comparable group of geometrically shaped formations were found on earth, geologists would quickly deem them artificial and archaeologists would organize an excavation to figure out how they were built. It's that simple.

The geometric diversity observed within the Conjoined Temple Complex, such as the circles, ovals, triangles, squares, and linear mounds (Fig. 3.14), can all be reproduced by utilizing the "Dome Arcology Model."

This entire group of geometric formations seen within the Conjoined Temple Complex could have been constructed as geodesic domes. They could have incorporated supportive armature of triangular, interlocking beams that are able to conform to any shape desired. Once built, they could be covered with concrete, metal, or other more exotic materials. Once abandoned these structures would eventually be covered with sand and dirt and over time they would appear as the earthen mounds produced by the indigenous people of North America.

Not far from the village of McFarland in Dane County, Wisconsin, is a group of earthworks known as the Sure Johnson Mound Group, which highly resembles the formations that occupy Tract 4 (Fig. 3.14). One section of the group has a circular mound and a long, linear, beam-shaped mound built on an elevated platform (Fig. 3.15 on page 86). The circular mound is about 20 feet across, while the linear mound is around 142 feet long.[3] The similarities between the two mound designs are truly stunning. They are almost identical.

A husband-and-wife team of archaeologists, Arlen F. Chase and Diane Z. Chase, recently utilized the state-of-the-art airborne LiDAR technology to take images of the ancient Maya site of Caracol. The site, which covers approximately seventy-seven square miles, is in what is now a section

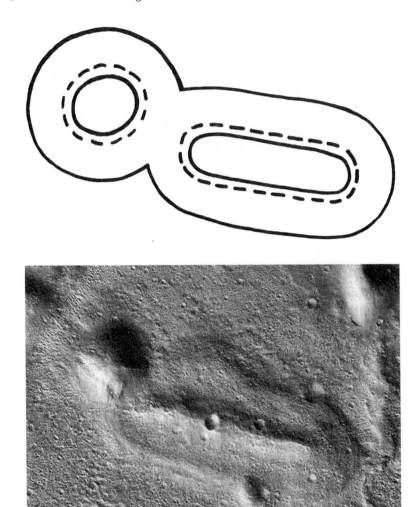

Fig. 3.15. Mounds on conjoined platform.
Top: Circular and linear mound. Detail of Sure Johnson Mound group.
Drawing by the author.
Bottom: Tract 4. Detail MRO HiRISE CTX
K11_057768_1854_XN_05N235W (2018).

of Belize.[4] When looking at the LiDAR image, which removes all the foliage, the site appears in a raw desert form, devoid of any plant life. With the dense canopy removed it reveals various buildings that sit on a massive set of connective platforms. Notice the pyramids, mounds, temples, and the long, bar-shaped structures.

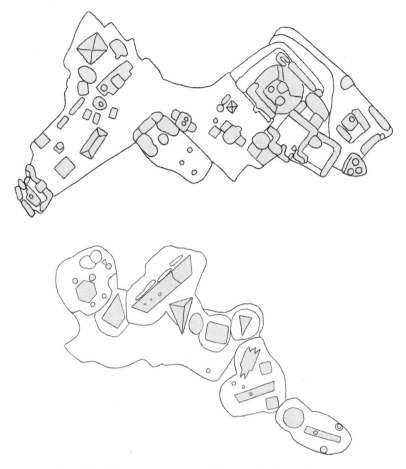

Fig. 3.16. Connective, undulating platform design.
Top: Caracol complex, Mexico (LiDAR).
Drawing and gray wash by the author.
Bottom: Conjoined Temple Complex. Nepenthes Mensae, Mars.
Drawing and gray wash by the author.

When the ruins of Caracol are compared to the formations observed on Mars at Nepenthes Mensae, their common positioning on large expansive platforms is quite revealing (Fig. 3.16).

Just as I have observed before: as above, so below.

FOUR

Elongated Hexagonal Mound

The Mars Revealer

BACK IN 1998, while I was a member of Richard Hoagland's Enterprise Mission discussion board, I met a very colorful and bombastic guy known as Gary Leggiere, who also uses the whimsical aliases Mad Martian and The Mars Reveler. Over the past thirty years Leggiere has proven to be one of the finest anomaly hunters to have ever scrutinized the surface of Mars. He has a discerning eye and a bewildering knack for finding faces and pyramids all over the planet and although he will openly share his finds, don't ever expect him to reveal their locations.

As an example, in October 2011 Leggiere posted a small portion of an ESA Mars Express image on his Facebook page that showed a large, elongated hexagonal formation[1] (Fig. 4.1). The image showed a highly symmetrical, six-sided formation that has a thick, relatively flat platform, which includes a central mound formation. There is also a small pyramidal formation that sits just beyond the edge of the platform on its lower eastern side.

After many weeks of multiple requests and intense negotiations as to the location of the formation, Leggiere finally broke down and provided data for the source image. He found the formation in the Nepenthes Mensae region of Mars from the ESA Mars Express image numbered H2004_0000_ND3 (Fig. 4.1). The image was acquired in the summer of 2005. It was taken in the early evening hours with a resolution of approximately twelve meters per pixel.

The Nepenthes Mensae region of Mars lies between 14°N and 0.45°S and expand from 99°E to 134°E.[2] The Elongated Hexagonal Mound is situated

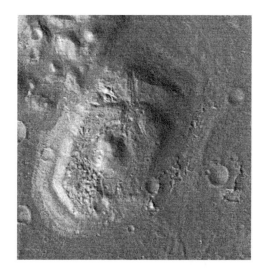

Fig. 4.1. Elongated Hexagonal Mound. Detail of ESA, Mars Express Orbiter H2004_0000_ND3 (2005). Image courtesy Gary Leggiere.

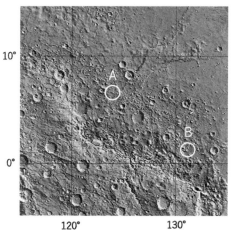

Fig. 4.2. Nepenthes Mensae region of Mars (MOLA map). Notated with the approximate location of the Mean City Complex at A and the Elongated Hexagonal Mound at B.

at approximately 2°N and 131°E, which much to my surprise, is only about 7°SE of the Mean City Complex, discussed in the previous chapter (Fig, 4.2). The MOLA map in Fig. 4.2 shows that location of the Mean City Complex, which is labeled A, while the location of the Elongated Hexagonal Mound is labeled B.

Themis and CTX

Going through the NASA archives for comparative images, I came across a second image of the Elongated Hexagonal Mound that was acquired by the Mars Odyssey THEMIS camera in January and released in June 2006. The THEMIS image shown in Fig. 4.3 on page 90 was taken during the winter

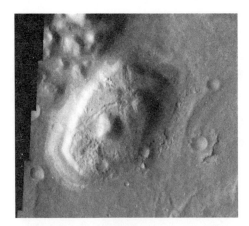

Fig. 4.3. Elongated Hexagonal Mound. Detail Mars Odyssey THEMIS V18218024 (2006).

Fig. 4.4. Elongated Hexagonal Mound. Detail MRO HiRISE CTX B03_010626_1821_XN_02N228W (2008).

in the late afternoon with a resolution of approximately 17 meters per pixel. Possibly because of the lower resolution the contours and surface detail in this THEMIS image are much softer. Notice the highly reflective western edge of the formation and lighting on the central mound shows it to be four-sided.

A third image of the Elongated Hexagonal Mound was acquired by the Mars Reconnaissance Orbiter HiRISE CTX camera in 2008 (Fig. 4.4). This image was taken in the winter during the midafternoon with the highest resolution of the group coming in at 5.3 meters per pixel.

Geological Analysis

During the latter part of 2011 I sent the three available images of the Elongated Hexagonal Mound to a geomorphologist William R. Saunders and a geologist Michael Dale for their review. They found a minimum of five different variations in appearance of surface materials over this small area of Nepenthes

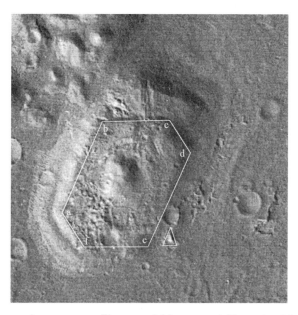

Fig. 4.5. Proposed geometry, Elongated Hexagonal Mound with geometrical contours overlaid by the author. Detail ESA H2004_0000_ND3 (2005).

Mensae. The hexagonal formation has six sides, and it has a square-shaped mound or hillock in the center (Fig. 4.5). The topography of the formation is relatively flat on its eastern side, while its western side is highly textured and scarred. The western side has what appears to be a grouping of structural foundations and the remnants of walls. Looking to the northern section of the formations platform it has a large, thick groove or fracture line that extends up in an almost northern direction. There is also a small, triangular mound sitting next to the eastern side of the hexagonal formation.[3]

If we examine the outlining contours of the Elongated Hexagonal Mound, we see it has six angles that appear to have equal opposing angles. Saunders and Dale suggest that its highly geometric form is the result of something not produced in nature. They noted that there is a network of pathways that appear from the center hill to the bottom in a rectangular pattern. The most western point (a) and the most eastern point (d) both have an angle of 120°, while the northwestern point (b) and the southeastern point have an angle of 115°. Comparing the northeastern point (c) with the opposing southwestern point (f), we see they both share an angle of 125°. As with any hexagonal formation the total sum of the internal angles of this formation adds up to 720°, the same total of angles achieved within a true hexagon.[4]

Fig. 4.6. Elongated
Hexagonal Formation.
Detail MOC M0903566
(1999).

Martian Comparisons

Looking to my personal collection of Martian anomalies, I recalled a similar elongated hexagonal formation in the Acheron Fossae region of the planet (Fig. 4.6). It also contains a central mound feature much like the Elongated Hexagonal Mound on Mars (Fig. 4.4). The image was obtained earlier by the Mars Global Surveyor camera back in 1999. The Mars Global Surveyor (MGS) spacecraft was launched from the Cape Canaveral Air Station in Florida in November 1996. Its mission was to study the entire Martian surface and to collect data about its atmosphere and topography. The spacecraft was equipped with a high resolution wide-angle camera system, known as the Mars Orbital Camera (MOC) that collected images to build a global map of the planet. The mission was officially ended in January 2007.[5] This MOC image was taken in the winter during the midafternoon with a remarkable resolution of 3.0 meters per pixels.

The formation is located about 28°N and 16°E of the Nepenthes Mensae region of the planet. It has an undulating hexagonal shape that has an elevated rim, which acts as a perimeter wall. The rim has an undulating waviness to its contours, as if it had been bent or partially melted. At the center of the enclosure there is an oval-shaped mound and a small, triangular form below it. There is an impact crater located to its west that has a high rim and shallow basin. There is also evidence of a thin ejecta blanket that flows out along the western wall of the Elongated Hexagonal Formation. This suggests that the impact event occurred after the Elongated Hexagonal Formation existed.

Fig. 4.7. Araban Hoyuk.
CORONA satellite
imagery (circa 1968).

Terrestrial Comparison

Located in the Araban Plain, just west of the Euphrates River in the southern region of central Turkey, are the remains of an elongated hexagonal mound that once supported the ancient settlement of Araban Hoyuk (Fig. 4.7). The settlement dates to the Neolithic period, around 8200 BCE.[6] Luckily for me, an aerial view of the area was acquired with the aid of the Central Intelligence Agency's CORONA satellite in 1968.[7] When I first found this image on the internet, I had a momentary loss of visual acuity—I squinted and rubbed my eyes and just stared at the screen in disbelief. It looked exactly like its Martian companion, a true doppelganger.

 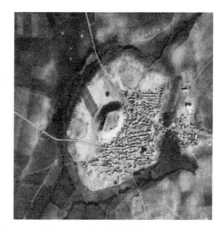

Fig. 4.8. Comparison of elongated hexagonal design.
Left: Elongated Hexagonal Mound.
Detail ESA H2004_0000_ND3 (2005).
Right: Araban Hoyuk, Turkey. Detail CORONA satellite imagery (1968).

Unfortunately, current satellite images of the formation show that the eastern and western sides of this ancient settlement are almost entirely covered over by a modern city.[8]

When the Elongated Hexagonal Mound on Mars is compared to the aerial view of the hexagonal mound at Araban Hoyuk, their common architectural design is truly uncanny (Fig. 4.8 on page 93). They both share the same elongated, hexagonal shape and include a central citadel. The two formations appear as mirror reflections of each other.

As I have seen so many times with these Martian structures, whatever kind of mound or pyramidal structure is observed on Mars, I am able to find similar ones on Earth. This one is most extraordinary. It's a facsimile of something that should not be there. Therefore, I must recommend that someone at NASA direct the Mars Reconnaissance Orbiter HiRISE camera to acquire additional higher resolution images of the Elongated Hexagonal Mound. They need to confirm its phenomenal geometric shape and acquire a closer look at its topography. And perhaps one day we can touch its broken stones and build our own encampment among its ruins.

FIVE

The Keyhole

A Wedge and Dome Formation

Exclamation Mark

ON JANUARY 11, 2011, the Mars Reconnaissance Orbiter (MRO) spacecraft acquired an image of something unusual on the surface of Libya Montes (Fig. 5.1). The onboard MRO HiRISE camera snapped an image of what appeared to be an odd wedge-shaped formation with an attached circular dome. The image was taken in the early afternoon with a resolution of fifty centimeters per pixel.[1] The official release on the University of Arizona website included a caption that accompanied the image, which referred to this odd, geometrically shaped formation as an "exclamation mark."[2] Traditionally, the basic shape of a conjoined wedge and dome formation are commonly referred to as a keyhole.

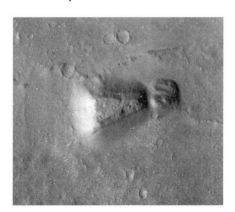

Fig. 5.1. Wedge and Dome formation. Detail MRO HiRISE ESP_020794_1860 (2011).

At the time of its release, I was totally unaware of this highly anomalous wedge and dome-shaped, keyhole structure. It was not until a good friend and colleague of mine at the Society for Planetary SETI Research, Greg Orme, brought this structure to my attention during July of 2013.[3] As soon as I downloaded the image I posted an article about it on The Cydonia Institute's discussion board tilted "Keyhole—Exclamation Mark on Mars" with a link to the original image.[4] Its reception was overwhelming, and the Keyhole quickly became the new hot topic on numerous YouTube videos and online news articles. Many of the reports used parts of my analysis and even my drawings without any mention of The Cydonia Institute or Greg Orme. The Keyhole structure on Mars was everywhere.

MRO and THEMIS

Excited with the discovery and all the attention it was getting, I performed an extensive search of the NASA archive, and I found two additional images of the keyhole structure that were taken three years earlier during the winter of 2007.

The first image of the Keyhole structure was acquired by the Mars Reconnaissance Orbiter (MRO) HiRISE spacecraft in November with its smaller CTX camera. The CTX image was taken during midmorning with a resolution of five pixels per meter (Fig. 5.2).

The second image of the Keyhole structure was taken by the Mars Odyssey THEMIS camera, which again captured the entire structure. This narrow-angle image was taken in December during the early afternoon, with a lower resolution of approximately seventeen meters per pixel (Fig. 5.3). The wedge and dome shape of the Keyhole structure is easily seen in both images, which are similar in tonality. The high sun is hitting the western side of the wedge

Fig. 5.2. Keyhole structure. Detail MRO HiRISE CTX P14_006672_1836_XN_03N267W (2007).

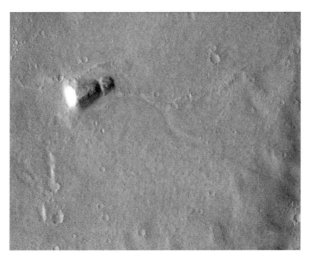

Fig. 5.3. Keyhole structure. Detail Mars Odyssey
THEMIS V26406033 (2007).

form, and the dark shadows give form to its southeastern side. The MRO
HiRISE CTX image shows the ribbed texture of the dome, and the sharp edge
of the wedge is more defined.

Libya Montes

The Libya Montes area of Mars is in the Syrtis Major hemisphere between
0.1° and 6.0°N and 80° and 96°E[5] (Fig. 5.4 on page 98). It is part of the
highly eroded and cratered remains of the southern rim of an ancient impact
basin called Isidis Planitia. Much of the region consists of networks of valleys
that run northward toward this large drainage basin. The geology of the area
appears to have been altered by an accumulation of strong wind action and
water erosion over time. This harsh weathering may have heavily degraded
this highland region and deposited sediment in its lowlands. The current
condition of this intriguing area has generated much debate among researchers in the scientific community as to how these highly modified and eroded
valley networks actually formed.[6]

During a 1999 study of this region, it was considered as a candidate for
the Mars Surveyor Lander, which was scheduled for launch on April 3, 2001.
Unfortunately, the mission was canceled in early 2000, and its spacecraft was
later used as the lander on the Phoenix mission, which landed in the north
polar region of Mars in 2008.[7]

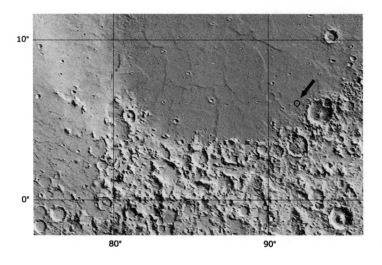

Fig. 5.4. Libya Montes. (MOLA data map). Notated with the approximate location of the wedge and dome formation (circle and arrow).
Image courtesy NASA/JPL/Malin Space Science Systems/The Cydonia Institute

Geology

Utilizing the available set of MRO and THEMIS images I was able to examine the defining aspects of its geometry and symmetry observed within the conjoined wedge and circular forms that make up the keyhole structure. A geological explanation of the formation is offered on the University of Arizona's website by Dr. Alfred S. McEwen, a Professor of Planetary Geology at Arizona State University.[8]

McEwen states "the origin of these hills may be difficult to understand on such ancient terrain. The straight edges suggest fractures related to faults. Maybe this feature was lifted by the faulting, maybe the surrounding terrain has been eroded down over billions of years, or both."[9]

I sent a copy of the image and the comments made by Dr. McEwen to two members of The Cydonia Institute, a geomorphologist William R. Saunders and a geologist Michael Dale, and asked them to provide their analysis of the Keyhole structure's geology and geometry. The first thing they noticed was that there was no visual evidence that this wedge and dome formation was thrust upward by faulting, as suggested by Dr. McEwen. In fact, they contend that the extended line running across the terrain is a fault line (Fig. 5.5) that runs between the wedge and the dome. If true, one side should be thrust up, the other side must be down. This is not the case, nor is their evidence of thrusting anywhere else along this line. If the features are glacial erratics, there remains

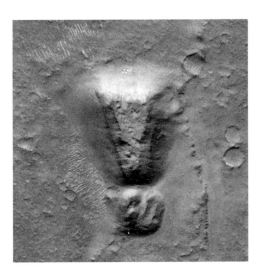

Fig. 5.5. Keyhole
Structure (Rotated 90°E).
Detail MRO HiRISE
ESP_020794_1860 (2011).

the question of why the wedge still has such prominent ridge lines while the tangent mound has been rounded off.[10]

Saunders conducted an expansive search of the surrounding terrain within the available data set, and he found no self-similar formations existed within or outside the region. It does not seem logical to have this wedge and dome formation totally isolated on an open plain, just sitting there, exposed to the same erosional forces to exhibit such dramatic geometrically opposing shapes. It appears as if this Keyhole structure was just placed there as it was.

I agreed with Saunders and told him that its isolated setting reminded me of an old car advertisement I had seen in a magazine. The ad showed a brand-new car sitting out in the middle of nowhere in an undisturbed wheat field. There were no dead bugs smashed on its headlights or any evidence of tire trails anywhere around it. It was as if the car was magically placed there, just as it was, as a finished product.

In examining the Keyhole structure Saunders noticed that dunes are formed around its perimeter in four different directions (Fig. 5.6 on page 100). The dunes observed on the western end of the wedge indicate wind from oppo-site directions. One would expect to see interference dune patterning where these opposing winds meet, but that is not the case here. It is also peculiar that there are no dunes or reworking on the apparently loose detritus material on the south side of the structure. There is also evidence of clustered pitting on the western edge of the wedge formation and within the dome formation (Fig. 5.6). This clustered pitting is also difficult to explain.

The pitting observed on the surface of the western side of the wedge

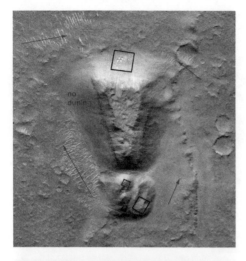

Fig. 5.6. Dunning and pitting on the Keyhole Structure (wedge and dome). Detail MRO HiRISE ESP_020794_1860 (2011). Notations by William R. Saunders.

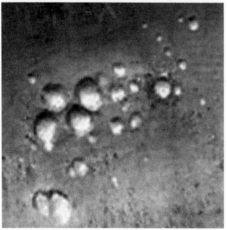

Fig. 5.7. Pitted and indented features on the Keyhole Structure (wedge and dome). Detail MRO HiRISE ESP_020794_1860 (2011).

formation appears as indentations within a hard-metallic surface or the remains of a hard surface material that has been gouged out or punctured by the impact of explosive projectiles.[11] Is the Keyhole structure made of some kind of metal or steellike material? These indents are smooth and circular in shape and look like the kind of ballistic indentations that you would see on the surface of a metal or aluminum plate after being shot with a high-powered gun (Fig. 5.7).

Geometry

Assuming there was extensive erosion over the surface of these conjoined formations, this pair of tangent formations exhibit remarkable symmetry. This symmetry goes against natural mechanisms. Symmetry, in nature, is found in biological systems, whereas inanimate objects exposed to environmental forces trend toward randomness and irregular abstraction. The Keyhole structure is approximately one and a half miles in length from the western side of the wedge form to the eastern edge of the dome form. It is approximately a half mile wide at its most western edge, from letter *a* to letter *c* in Fig. 5.8, while the dome formation is approximately 2,500 feet in diameter.

The geometry of the Keyhole structure represents two very basic geometries: the 360° circle and the 180° isosceles triangle (Fig. 5.8). The large dome-shaped formation, located directly below the wedge-shaped formation, appears almost circular, while the overall shape of the wedge formation takes on the form of a trapezoid with an isosceles triangle on its uppermost surface. The angles of this triangle are *x* and *y* = 75° and *z* = 30°, adding up to 180° (Fig. 5.8).

The line *i* drawn to bisect ridgeline *a* to point *c* also bisects lines from points *e* and *f*, along with points *g* and *h* and points *b* and *d* demonstrating the bilateral symmetry to the wedge or trapezoid section. The upper and lower ridges of the wedge appear as nearly perfectly straight lines running parallel to each other (Fig. 5.8). The lower ridge line extending from point *a* to point *b* is parallel to the upper ridge line that extends from point *e* to point *g*. The adjoining upper ridge line, extending from point *f* to point *h* is parallel with the lower ridge line extending from *c* to point *d*.[12]

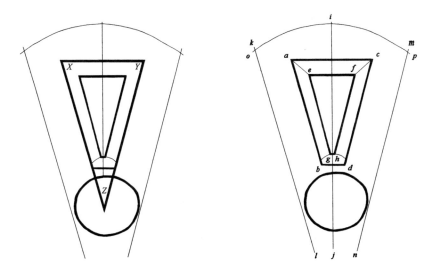

Fig. 5.8. Keyhole Structure (wedge and dome) geometry. Analytical drawings by the author. Mars Reconnaissance Orbiter ESP_020794_1860 (2011).
Left: Triangular section. Notated X, Y, and Z.
Right: Wedge section. Notated a–p.

Terrestrial Comparison

There is a long history of worldwide cultures utilizing the keyhole shape to produce a variety of graphic-based designs and monumental structures. As a symbol it is often related to death and the afterlife. Starting with the Al Hait and Khaybar regions of Saudi Arabia, there are dozens of keyhole-shaped tombs that were discovered in 2009 with the aid of satellite images provided by Google Earth[13] (Fig. 5.9). Thought to be over six thousand years old[14] the main body of these tombs takes the shape of an isosceles triangle with a circular mound attached at its vertex. The keyhole-shaped tombs are commonly aligned with the rising sun.[15]

Moving westward over the Red Sea and into the deserts of Egypt we enter the Theban necropolis at Deir el-Medina. It is here that we find the tomb of Amenemipet that contains a mural depicting the goddess Hathor as a cow (Fig. 5.10). She sits with her body bound in wrappings and the sun disc of Ra rests between her horns. On her back are a flail, which is a symbol of the Pharaoh, and a large, keyhole-shaped menat, which is a symbol of Hathor. Priestesses would wear the menat as a necklace for good fortune and to protect them against evil spirits. It was also placed with the dead to protect them in the afterlife.[16]

Traveling across the Atlantic Ocean to the other side of the planet, there are earthen mounds that take on the shape of wedge and dome formations that were produced by the early Mound Builders of North America. One of these wedge and dome keyhole-shaped formations is situated on a ridge along the eastern side of Lake Monona, near the city of Madison, Wisconsin (Fig. 5.11). The oddly shaped mound is referred to by archaeologists as a "war

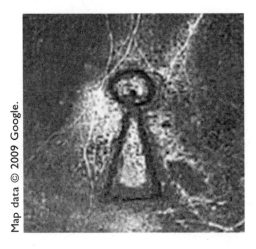

Map data © 2009 Google.

Fig. 5.9.
Keyhole tomb.
Saudi Arabia,
4000–1000 BCE.

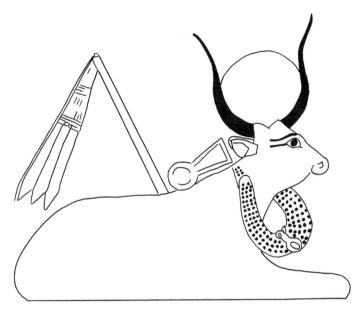

Fig. 5.10. Hathor with menat (keyhole).
Detail of mural at Deir el-Medina, Egypt (Nineteenth Dynasty).
Drawing by the author.

club." The circular, dome-shaped section of the formation is 36 feet across, while the extending wedge form is 165 feet long, measuring 20 feet wide at its widest part.[17]

The Pre-Columbian cultures of Mexico also produced structures and graphic images that adhered to the keyhole design. The first example is carved on a structure at the city of Uxmal in the Yucatan (Fig. 5.12). Its glyphic form features a feathered rosette carved in the shape of a keyhole with a six-pointed hexagram in

Fig. 5.11. Wedge and
dome (war club) mound.
Lake Monona, Wisconsin.
Drawing by the author.

Fig. 5.12. Pre-Columbian keyhole designs.
Left: Feathered rosette. Uxmal, Yucatan (1000 BCE).
Drawing by the author based on *Ancient Past of Mexico* by Alma Reed, page 12.
Right: Keyhole symbol. Detail of polychrome vessel. Veracruz, Yucatan.
Drawing by the author after Justin Kerr K7006.

its center.[18] The second example is from Veracruz, Mexico. It is here that we find a keyhole symbol painted on the side of a polychrome vessel (Fig. 5.12).

The third example is found within the architecture of the Maya site of Palenque, which is in the state of Chiapas in Mexico. Within this temple

Fig. 5.13. Temple of the Foliated Cross. Palenque, Mexico, 684 CE.
Notice the keyhole-shaped openings that flank the central
triangular opening. Drawing by the author.

Fig. 5.14. Keyhole by Vik Muniz, 2006 (Carajás N4, Iron Mine).

complex there is a structure known as the Temple of the Foliated Cross (Fig. 5.13). On the outer facade of the temple is a central, triangular arch that is flanked by a pair of keyhole-shaped openings. Being symbols of death, perhaps they are providing paths to the Otherworld.

The elegant form of this keyhole design extends well into contemporary art projects such as the geoglyphic creations of Brazilian land artist Vik Muniz (Fig. 5.14). Inspired by the design of classic symbols and icons used in popular culture in 2006, Muniz produced a set of geoglyphs on a large plot of land owned by the Carajás Mining Complex in Northern Brazil.[19] One of the landforms he produced on this abused and barren landscape is the outline of a traditional keyhole formation (Fig. 5.14).

The most impressive example of a keyhole-shaped structure produced on Earth is found in Kofun, Japan (Fig. 5.15 on page 106). This well-preserved tomb is an earthen mound that is thought to have been built in the fifth century. It measures 120 meters long.[20] This classic wedge-and-dome-shaped tomb is highly comparable to the Keyhole-shaped formation observed on Mars.

While examining these ancient keyhole-shaped tombs found in Japan, Dr. Robert Schoch, a professor of Natural Science at the College of General Studies at Boston University, stated that he believes that many of these ancient keyhole-shaped tombs may have been created as geoglyphic markers, signaling the attention of aerial travelers to "stop and look here."[21]

Fig. 5.15. Keyhole tomb. Kofun, Japan, circa 400 CE. Drawing by the author.

The structural design of the keyhole formation on Mars remains exceptional in regard to its geometry and symmetry. The continuity of architectural references is eloquently expressed within the monuments produced by New World, Middle Eastern, and Japanese cultures where a common aesthetic is strongly supported.

The conjoined wedge and dome formations observed in Libya Montes are well proportioned and highly symmetrical despite the actions of natural depositional and erosional agents. While there are known geological mechanisms that are capable of creating and destroying the individual angles and planes presented in this formation, the natural creation of two opposing geometrically designed formations seems to go well beyond the probability of chance.

As of this publication the Keyhole structure has appeared in various YouTube videos, newspapers, and magazine articles. It was also the subject of a science paper in the *Journal of Space Exploration* and was a featured topic on the History Channel's *Ancient Aliens* and *The UnXplained* programs. Since its discovery by NASA, it has become clear that this Keyhole structure holds the public's interest and provides compelling support for a broader investigation by independent planetary scientists.

SIX

The Martian Atlantis I
Twin Cities in Chaos

A Grid of Cellular Formations

ON JULY 21, 2019, Greg Orme of the Society for Planetary SETI Research posted a section of a Mars Reconnaissance Orbiter HiRISE image on his Facebook page, titled "The Kings Valley, Mars, Why We Must Go."[1] Although the image was over nine years old this was the first time, to my knowledge, that anyone noticed this area of Atlantis Chaos contained the remains of a tightly knit grid of cellular formations (Fig. 6.1).

Image courtesy Greg Orme.

Fig. 6.1. Section of Atlantis Chaos.
Detail MRO HiRISE ESP_019103_1460 (2010).

Fig. 6.2. The City Complex.
Detail MRO HiRISE ESP_019103_1460 (2010).
Notice the evenly spaced grid and blocky mounds.

The same section of the MRO HiRISE image was posted a week later, on another Facebook page (Fig. 6.2) of an independent researcher known as Javed Raza.* He gave the area a closer look and began highlighting some of the linear formations that appeared to be part of a massive city-like complex. He saw the remains of what he thought were intelligently made structures that were built in a "grid pattern." Raza suggested that the arrangement of these evenly spaced foundations with broken walls and towers are typical of the kind of remains one would see in built-up areas on Earth.[2]

The MRO HiRISE image in Fig. 6.3 was acquired in the summer of 2010 during the early afternoon with a resolution of fifty centimeters per pixel. The area of interest is positioned just to the west of a pair of joined craters.

If I zoom out from the cellular formations of the city I can get a contextual view of the Martian Atlantis area observed in the MRO HiRISE image shown in Fig. 6.4. The cropped image shows a pair of large-impact craters in the upper right-hand corner and the location of the cellular formations are circled just to the west of them, which is labeled A in Fig. 6.3. Notice the fine meshwork of the highly reflective grid-like pattern running across the surface within the circled area.

---

*Javed Raza also uses the Facebook name Jay Raza.

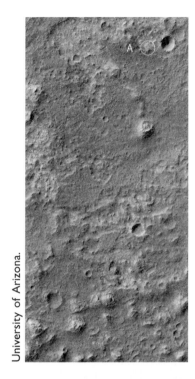

University of Arizona.

Fig. 6.3. The Martian Atlantis. Detail MRO HiRISE ESP_019103_1460 (2010). The location of the City Complex is circled and labeled A.

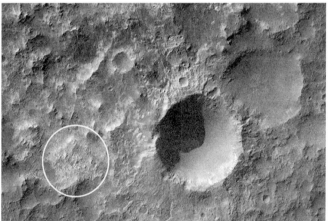

Fig. 6.4. The City Complex, labeled A in Fig. 6.3.
Detail MRO HiRISE ESP_019103_1460 (2010).
The location of the City Complex is circled.

The City Complex

Soon after Javed Raza exposed the remains of his Martian Atlantis to a wider audience on Facebook, an image analyst and processor, Neville Thompson,

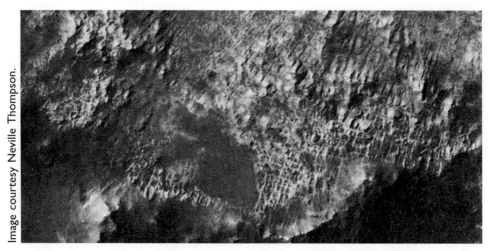

Image courtesy Neville Thompson.

Fig. 6.5. The City Complex. Detail MRO HiRISE ESP_019103_1460 (2010).

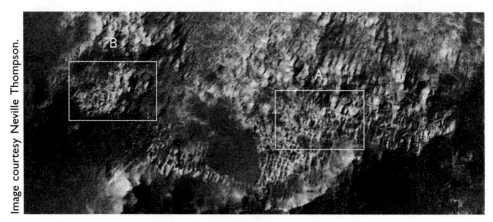

Image courtesy Neville Thompson.

Fig. 6.6. The Twin City Complex.
A: The Eastern City Complex. B: The Western City Complex.
Detail MRO HiRISE ESP_019103_1460 (2010). Notated by the author.

released his large Giga-pan of the area on August 7, 2019. Utilizing the same MRO HIRISE image (ESP 019103 1460) Thompson's version of the image included an expansive view of the entire blocky terrain surrounding the main City Complex in much more detail[3] (Fig. 6.5). The format of the GigaPan created by Thompson is very user friendly. It allows the viewer to zoom in and out of the image and focus on the various remnants of what appear to be buildings and walls.

The City Complex can be divided up into two sections that exhibit a high concentration of grid patterns of broken walls and blocky mounds that form two separate cities. There is a cluster of grid patterns on the left side and a larger cluster on the right. The first of these "twin" cities is the original City Complex that was reported by Orme and Raza, which I have titled the Eastern City Complex (labeled A in Fig. 6.6). The second city, which is located on the western side of the complex, I have titled the Western City Complex (labeled B in Fig. 6.6).

The Eastern City Complex

Utilizing Thompson's GigaPan I moved in closer to get a better look at the highly reflective gridwork observed within the Eastern City Complex (labeled A in Fig. 6.6). The entire area is filled with a mesh of linear walls and cubic structures that sit among the ruined foundations of decayed or destroyed buildings and exposed mounds (Fig. 6.7). The Eastern City Complex measures approximately three hundred meters from its western side to its eastern edge, while each of the cubic cells are about ten meters square.[4]

Starting with the tall tower-like structure (labeled A in Fig. 6.8 on page 112), it appears to be the remains of a pyramidal temple with a steep incline that I have titled the Twisted Temple. The second structure (labeled B in Fig. 6.8) has a broad dome shape with a spiked peak that I have titled the Spiked Dome. The third structure (labeled C in Fig. 6.8) is located on the far right side of the

Image courtesy Neville Thompson.

Fig. 6.7. The Eastern City Complex.
Detail MRO HiRISE ESP_019103_1460 (2010).

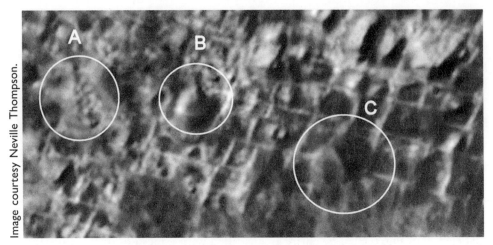

Image courtesy Neville Thompson.

Fig. 6.8. The Eastern City Complex.
A: Twisted Temple. B: Spiked Dome. C: Mill Factory.
Detail MRO HiRISE ESP_019103_1460 (2010).
Contrast adjusted and circles added by the author.

Spiked Dome in the lower right side of the complex. The formation looks like a partially mud-covered structure that resembles a barn or an old mill factory.

Atlantis Chaos

The Atlantis Chaos region of Mars is found in the southwestern quadrangle of the planet within a lowland plain, on the western side of Terra Sirenum. The area contains a few hundred small peaks and buttes interspersed with flat-topped hills known as mesas.[5] It is located within 34.7°S and 177.6°W[6] (Fig. 6.9). The area of interest for this review is found on the eastern side of a large circular imprint on the surface, highlighted by a black circle in Fig. 6.9. Scientists believe the circular imprint is the remains of a large lake.[7] The presence of a lake would provide the perfect environment to establish a port settlement along its edge.

This area, referred to as Atlantis Chaos, finds its origins in Greek mythology.[8] As we all know, Atlantis is a mythical place that was written about by the Greek philosopher Plato. He told a tale of a highly advanced civilization that lived on a circular island that disappeared in a worldwide cataclysm. The word *chaos* comes from Greek creation stories that tell us, in the beginning, the universe was filled with Darkness, and from that emptiness sprang Chaos and therefore everything that occupies the universe sprang from Chaos.[9]

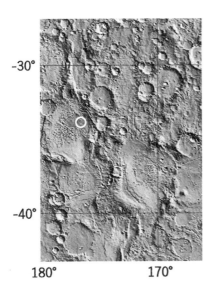

Fig. 6.9. Atlantis Chaos region of Mars (MOLA data map). Notated with the location of The Martian Atlantis (circle).

Terrestrial Comparison

Looking at the overall compartmentalized units that blanket the Eastern City Complex, one can see their cubic grid design takes on the form of mudbrick and stone adobe houses built throughout the Midwest of the United States. A great example is in located within the northwestern region of New Mexico known as the Aztec Ruins National Monument (Fig. 6.10). Although the original structures were thought to have been built by the Aztec, the site is now attributed to the early Pueblo people and dates to around the twelfth century.[10] Notice the fine grid work and cellular rooms are very similar to the formations observed within the ruins at Atlantis Chaos on Mars.

Fig. 6.10. Pueblo ruins, Aztec Ruin National Monument. New Mexico, twelfth century.

Map data © 2020 Google.

Map data © 2015 Google.

Fig. 6.11 Al-'Ula (aerial view), Saudi Arabia.

Similar boxlike mud-and-brick dwellings are also found on the other side of the world at Al-'Ula in Saudi Arabia, which share a common templet for a cubic design (Fig. 6.11).

In Saudi Arabia, there are the remains of a once bustling city known as Al-'Ula. Built over two thousand years ago it was a major hub of commerce and industry. It was located right along the old "Incense Road," which was a network of routes that facilitated the trading of spices, silk, and other luxury items through Arabia, Egypt, and India. The city remained occupied up until the 1980s, however much of the condensed city now lies in ruins.[11] Notice the ragged, broken walls and missing roof tops. This grid of hollow dwellings in Saudi Arabia looks a lot like the meshwork of cellular formations that we are seeing on Mars (Fig. 6.12).

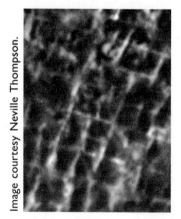

Image courtesy Neville Thompson.

Map data © 2015 Google.

Fig. 6.12. Cubic walls.
Left: Section of the Eastern City Complex, Mars.
Detail MRO HiRISE ESP_019103_1460 (2010).
Contrast adjusted by the author.
Right: Al-'Ula, Saudi Arabia.

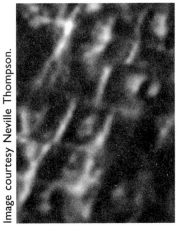

Detail of image by Evan Sanchez.

Image courtesy Neville Thompson.

Fig. 6.13 Peaked wall design.
Left: Ruins of urban buildings, Machu Picchu.
Right: Section of the City Complex, Mars. Detail MRO HiRISE
ESP_019103_1460 (2010). Contrast adjusted by the author.

Another comparable boxlike architectural design can be seen in the rows of urban buildings that are on the hill sides of Machu Picchu, which rest on a flat mountaintop high above the Sacred Valley of Peru (Fig. 6.13). Scattered among its main structures and outlying buildings are rows of smaller dwellings that were constructed out of polished, dry stone walls.[12]

When these rows of smaller dwellings are compared to a section of the Eastern City Complex that sits on the eastern side of the Spiked Dome structure, their common cubic design becomes almost indistinguishable (Fig. 6.13). Notice the conjoined parallel walls and peaked roof supports. If these boxlike structures, which we are seeing on Mars were found on Earth, archaeologists would surely see them as the remains of ruined adobes.

The Temple, Dome, and Mill Factory

The three main structures that have been recognized in the Eastern City Complex—the Twisted Temple, the Spiked Dome, and the Mill Factory (labeled A, B, and C in Fig. 6.9)—can be compared to terrestrial buildings.

I'll start with the structure that I refer to as the Twisted Temple (labeled A in Fig. 6.9); its steep pyramidal form has an undulating, segmented shape that is topped with a circular knob (Fig. 6.14). The stepped facade on its eastern side appears as a cascade of twisted, tubular blocks and

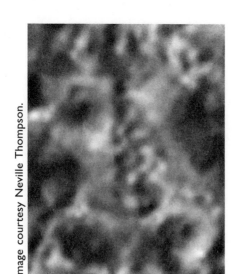
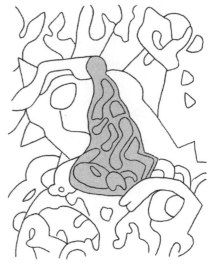

Image courtesy Neville Thompson.

Fig. 6.14. The Twisted Temple.
Left: Detail MRO HiRISE ESP_019103_1460 (2010).
Right: Analytical drawing and gray wash by the author.

pillars that have been mangled together. Its base sits on a wide drum-shaped platform within a pitted moat that includes a large polygonal formation on its eastern side. The Twisted Temple is the tallest structure in this area of the complex. It measures approximately sixty-four feet high, which is about the height of a six-story building.[13] An analytical drawing is provided in Fig. 6.14.

Raza was one of the first to notice the Twisted Temple structure,[14] and he suggested that its segmented form could be compared to the Virupaksha Temple, which is located in the state of Karnataka, India (Fig. 6.15). The Virupaksha Temple is a beautifully designed work of Indian architecture. The structure has repeated patterns that divide and repeat themselves over and over all the way from its base to its peak. The temple is over 160 feet high and has a steep, nine-tiered design that is topped by a small shrine.[15]

Although I initially found Raza's comparison to have merit, after taking a long, hard look at the interweaving design of the Twisted Temple on Mars I concluded that it had no equal on Earth, until I came across the twisting sky-scrapers of Belgian ecological architect Vincent Callebaut (Fig. 6.15).

As soon as I saw his Agora Tower in Taipei, Taiwan (Fig. 15), I thought I was seeing a doppelgänger of the Twisted Temple on Mars. Callebaut's twisted tower was inspired by the double heliacal structure of DNA. The tower's

Fig. 6.15. Temple comparison.
Left: Virupaksha Temple. Karnataka, India, circa seventh century.
Right: Angora Tower. Taipei, Taiwan. Vincent Callebaut Architectures (2016).
Drawings by the author.

ecofriendly design has suspended gardens that wrap around its balconies, allow-ing residents to harvest their own food.[16]

The second formation to be examined is the Spiked Dome structure (labeled B in Fig. 6.9). It has a very distinct dome or beehive shape that is topped with a tall spire. It rises above its surrounding cubic building by about fifty feet.[17] An analytical drawing is provided in Fig. 6.16 on page 118.

The overall form of the Spiked Dome on Mars strongly resembles the dome-shaped stupas that were commonly constructed throughout India and Sri Lanka. A great comparison can be made with the similar form of the Rankoth Vehera stupa that was built in Sri Lanka in 1190 CE (Fig. 6.17 on page 118). Made entirely of brick, it sits on a square platform and has a height of almost two hundred feet.[18]

Because I live in the United States, the first thing I thought of when seeing the Spiked Dome structure on Mars was that it looks a lot like the domed roof that crowns the U.S. Capitol building in Washington, DC (Fig. 6.17). Notice how both structures have a broad circular base and domed arch topped with a spire at its summit. Their spiked-dome designs are almost identical.

I have named the third formation of interest in this section of the Eastern City Complex the Mill Factory (labeled C in Fig. 6.6) because it resembles a large barn or lumber mill. This large cubic structure is rectangular in shape and has a large docking porch attached to its southwestern side (Fig. 6.18). The

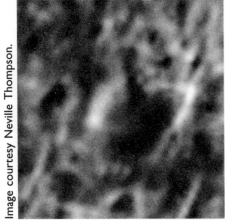
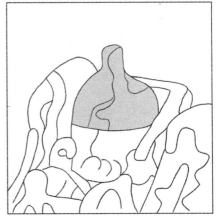

Fig. 6.16. Spiked Dome.
Left: Detail MRO HiRISE ESP_019103_1460 (2010).
Right: Analytical drawing and gray wash by the author.

Fig. 6.17. Dome comparison.
Left: Rankoth Vehera stupa. Sri Lanka, 1190 CE.
Right: Capitol building. Washington, DC.
Drawings by the author.

main building has a slanted roof and an upper loft or chimney that is approximately sixty-one feet high.[19]

The Mill Factory is supported by a rocky foundation that appears to rest on the shoreline of a now-dry waterfront. The cubic-shaped building sits near an outcrop of bound stones that may have faced a great river. If this is just the remains of a mud mound and accumulated debris, it certainly has taken on the

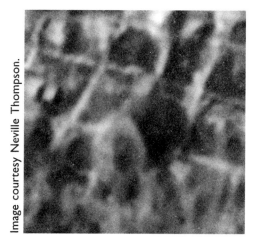
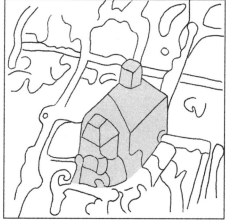

Image courtesy Neville Thompson.

Fig. 6.18. Mill Factory.
Left: Detail MRO HiRISE ESP_019103_1460 (2010).
Right: Analytical drawing with gray wash by the author.

appearance of a sculpted mound that looks a lot like a large barn or possibly a lumber mill. An analytical drawing is offered in Fig. 6.18.

While doing a little research into the construction and design of old barns, factories, and lumber mills, I quickly came across a terrestrial example in Jericho, Vermont. Known as the Old Red Mill, it sits along the Brown River on a strong fieldstone foundation. It has a gabled roof and a large square tower on its northeastern side. It has a wheelhouse that projects into the river and a stone retaining wall.[20] Its overall design highly resembles the Mill Factory observed on Mars (Fig. 6.19).

Image courtesy Amber Ramhorn.

Fig. 6.19. Old Red Mill. Jericho, Vermont, circa 1856.

Geological Comparison

After spending many weeks examining these compartmentalized cellular formations, geomorphologist William R. Saunders was at a loss to explain them. He could not find any natural geological process on Earth that could create the types of formations he was seeing within the Eastern City Complex on Mars. None of them would create a cluster of enclosed, boxlike formations that are hollow with high standing walls.

Included in his study Saunders offered a few natural geological processes that can create grid-like formations, such as tessellated pavements, karst formations, and honeycomb weathering.[21]

Although similar, as is evident in their topographical imprint, none of these natural geological formations exhibit the same boxlike features with high standing walls that we see on Mars.

The first example of a similar geological process offered by Saunders is a tessellated pavement found on the Tasman Peninsula in Tasmania. Tessellated pavements are naturally created grid patterns of square and polygonal blocks that resemble mosaic tiles (Fig. 6.20). Unlike the deep and hollow boxlike structures with high standing walls observed on Mars, the profile of tessellated pavements is relatively shallow. They occur in flat areas, and their borders can

Photograph courtesy Robert David Siegel, M.D., Ph.D., Stanford University.

Fig. 6.20. Tessellated pavement.
Tasman Peninsula, Tasmania, 2015.

Fig. 6.21. Karst formations.
Tsingy de Bemaraha, Madagascar, 2016.

appear as ruts or groves, while others have slightly raised borders that are generally uniform in shape. They are often formed by the cracking and lithification
of mud on the top of layered beds of sandstones and other sedimentary rock.[22]

The second example is what geologists call a karst. A great example can
be found in the northwestern part of Madagascar in the Tsingy de Bemaraha
National Park (Fig. 6.21). This type of erosional pattern appears as a spiked or
fluted rock outcrop. These formations are created in coarsely fractured rocks
where there is enough running water that causes the dissolution of soluble
rocks such as limestone, dolomite, and gypsum. They are created by the erosion caused by water drainage where the harder, more weather-resistant rock
remains as the lesser weather-resistant rock around it dissolves, leaving a pattern of raised, fluted stone.[23]

Karst features are in full opposition to the gridded cell-like structures
found on Mars. The jagged, pointed profile of a karst formation is carved out
of the surrounding rock, exposing a pitted, sharp textured stone, while the cellular formations on Mars are empty boxlike enclosures with free standing walls.

The third example offered is the honeycomb and lacework weathering
patterns often found within sandstone and basalt. It is a form of cavernous

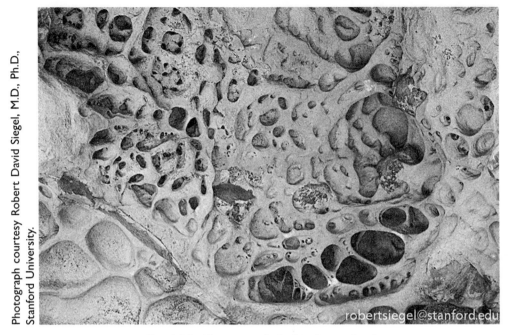

Fig. 6.22. Honeycomb (tafone) weathering.
Castle Rock State Park, California, 2014.

weathering that consists of regular, tightly adjoining cavities that are commonly small and around an inch in size. A great example is at Castle Rock State Park in California (Fig. 6.22). Notice the pitted texture of cellular crevasses that are created by the repeated exposure of salt water. The sea water penetrates the porous rock cracks and as it evaporates, salt crystals are formed inside. This creates pressure within the rock that is released by breaking off bits and pieces of rock. As this erosion effect is repeated over and over these honeycomb-shaped holes are created.[24]

Of the three examples offered by Saunders it is the honeycomb weathering pattern that comes the closest to the type of cellular features we are seeing on Mars. However, honeycombs tend to be erratic in shape and have rounded corners. They are not as large or as square and uniform as what we are seeing in the boxlike formations within the Eastern City Complex at Atlantis Chaos.

The City of Two Faces

On October 12, 2019, an independent researcher, Ennio Piccaluga, announced that he had found a sculptural portrait of a human face looking up from the

 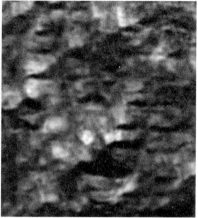

Fig. 6.23. The Facial Mask.
Left: Eastern City Complex with location of face (circled).
Detail MRO HiRISE ESP_019103_1460 (2010).
Right: Detail of Face Mask (rotated to the right).
Detail MRO HiRISE ESP_019103_1460 (2010).

Eastern City Complex[25] (Fig. 6.23). The facial features include two eyes, a nose, full cheeks, and slightly puckered lips above a full chin. The overall U shape of the face is flanked by rectangular blocks that highlight its U shape, while a distinct straight line demarks the top of the forehead, like a lintel. The facial mask is about thirty meters wide and fifty meters in length, including chin and forehead. That's about ninety-eight feet wide and one hundred sixty-four feet in length. To get a better perspective on its size, the head of George Washington at Mount Rushmore in South Dakota is only sixty feet tall.[26]

The facial features observed within the Facial Mask, found near the Eastern City Complex, appear to exhibit Mesoamerican motifs. Here is a comparison of the Eastern City Complex Facial Mask with a Pre-Columbian stone mask (Fig. 6.24 on page 124). Notice both masks have a flat forehead, two eyes, a broad nose and slightly parted mouth. Each mask also has a set of horizontal "flayed" bands that run across the face like a mummy's wrapping.

Stepping back from the Facial Mask that is set within the northern grid pattern of the Eastern City Complex, I noticed that the entire area of the complex conforms to the shape of a half-faced portrait of a human face with a winged headdress (Fig. 6.25 on page 124).

When this large half-faced portrait is cropped along the proposed demarcation line, which can be seen along the partial mouth and nose, another

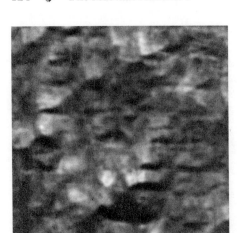 

Fig. 6.24. Mask Comparison.
Left: Facial Mask. Detail MRO HiRISE ESP_019103_1460 (2010).
Right: Flayed Face (stone mask). Teotihuacan, third century CE.
The De Agostini Collection. Drawing by the author.

Image courtesy Neville Thompson.

Fig. 6.25. The Eastern City Complex, Half Face.
Detail MRO HiRISE ESP_019103_1460 (2010).
Demarcation line added by the author.

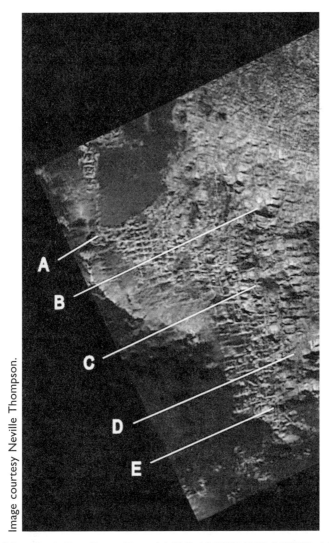

Fig. 6.26 The Grid City Face. Detail MRO HiRISE ESP_019103_1460 (2010).
Gray wash and notations A–E by the author.

Mesoamerican-styled portrait is revealed (Fig. 6.26). Notice the large, winged headdress within the gridded landscape (labeled A in Fig. 6.26). The small Facial Mask is embedded within the forehead (labeled B in Fig. 6.26), and directly below that is the imprint of an almond-shaped eye (labeled C in Fig. 6.26). The nose bridge begins just past the inner corner of the eye and leads down to an elevated, highly ornamented nose with nostrils (labeled D in Fig. 6.26). Just below that is a snarling mouth with lips and upper and lower fangs (labeled E in Fig. 6.26).

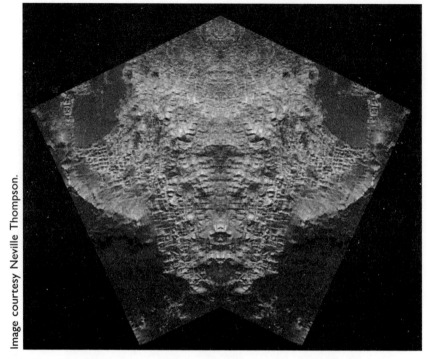

Image courtesy Neville Thompson.

Fig. 6.27. The City Grid Face. Duplicated detail of MRO HiRISE
ESP_019103_1460 (2010). Duplication and gray wash by the author.

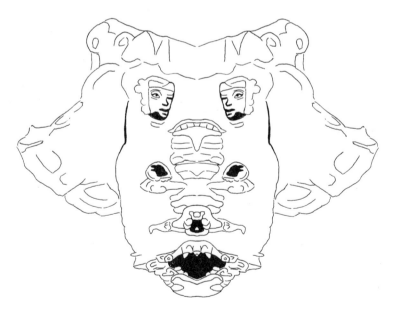

Fig. 6.28. The City Grid Face. Analytical drawing by the author.
Duplicated detail of MRO HiRISE ESP_019103_1460 (2010).
Image courtesy Neville Thompson.

When the half-faced profile within the city is duplicated along the extent of its projected demarcation line, a full-frontal view of a Mesoamerican-styled portrait is revealed (Fig. 6.27). An analytical drawing of the City Grid Face is provided in Fig. 6.28.

Terrestrial Comparison

Designing and planning a city in the shape of a human face may seem extraordinary and impractical; however, just such a city was built in Buenos Aires, Argentina in 1947. The city, known as Evita City, was commissioned by Argentine President Juan Perón to conform to the shape of his wife Evita Perón's profile[27] (Fig. 6.29). The city was constructed right next to the airport, so visitors coming to Buenos Aires would be greeted by her face.

The silhouetted, profiled face follows the contours of a female head with a sloping nose, firm chin, and strong jaw line. It also has a thick neck and her signature hair bun. The interior features of the head display a tight gridwork of streets and houses that adhere to a set of compartmentalized facial features with an open parcel of land that acts as an eye. Its overall design resembles a high-tech circuit board, which is not very flattering as a female portrait.

In contrast to the cut-in-half design of the City Grid Face on Mars, the simple profiled design of the Evita City is easier to see and identified as a

Fig. 6.29. Evita City. Outlined contours added by the author.

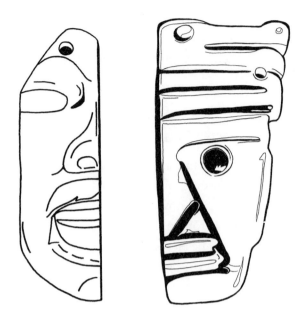

Fig. 6.30. Cut-in-half faces.
Left: Olmec jade pendant. Drawing by the author after a photograph
by Hector Gambora P.
Right: Maya jade amulet. Costa Rica National Museum.
Drawing by the author.

face, but lacks the detail observed in its Martian counterpart. Although it is portrayed as a half face, the City Grid Face on Mars has a well-defined eye, nose, and mouth with fangs that stand alone before the image is duplicated.

This idea of producing half-faced sculptures finds its origins in New World cultures, such as the Olmec, Maya, Aztec, and even the Indian cultures of North America. Here are two great examples: The first is an Olmec pendant, which was found in Bagaces, Costa Rica, that is in the shape of a half face[28] (Fig. 6.30). Notice this finely carved jadeite pendant has been cut perfectly in half, from the top of the forehead right down through the nose and mouth. A second example was produced by the Maya, which shows a human head with a band headdress that is topped with the profiled head of a crocodile. Like the Olmec example, the face was cut entirely in half. Notice the human face has a circular eye, an elongated, rectangular ear, a triangular nose, and a pair of slightly parted lips (Fig. 6.30).

It is believed by archaeologists that these "cut in half" masks and figurative objects were originally produced as complete sculptures that were ceremonially cut in half. After a king or member of the elite family died some of his personal objects were collected and cut in half and then placed in the grave as burial offerings. One half of the object was placed in the grave with the deceased, to accompany him in the Underworld, and the other half was kept in the Upperworld and given to another member of the elite or a family member.[29] With the separation of these personal objects the essence of the deceased would remain in both worlds.

Although being physically separated these individual halves were considered complete representations of the original whole object. Each piece of the object was seen as an individual element of a holographic whole that retains the entire object. This is a Mesoamerican concept known as *pars pro toto,* which enables any part of an object to be used as a representation of the whole.[30] Therefore, this act of cutting the object in half embraces the duality of life and death. Each half creates two equal parts of the whole that represent a mirrored reflection of two opposing worlds: one side embodies the living force of the Upperworld, while the other side represents the soul's descent into the realm of the Underworld. Perhaps the half-faced portrait observed within the overall design of the Grid City Face at Atlantis Chaos is another example of pars pro toto, or "as above so below."

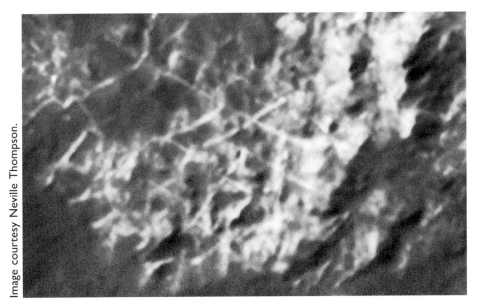

Image courtesy Neville Thompson.

Fig. 6.31. The City Complex (west side).
Detail MRO HiRISE ESP_019103_1460 (2010).

The Western City Complex

On the far western side of the Eastern City Complex, beyond the smooth sand and silt-filled terrain, is another area showing rows of gridwork that I have titled the Western City Complex (labeled B in Fig. 6.7). A close-up of the area from Neville Thompson's enhancement in Fig. 6.31 on page 129 shows additional walls and cellular structures. Notice the massive gridwork of hollow buildings and ruined foundations with standing walls that appear to be built

Detail of photograph by William Vandivert/
The LIFE Picture Collection/Shutterstock.

Fig. 6.32. Bombed-out building comparison.
Top: The City Complex (West side).
Detail MRO HiRISE ESP_019103_1460 (2010).
Bottom: Aerial view of damaged buildings during the
Battle of Berlin in Germany, July 1945.

on an elevated terrace. If these formations were observed here on Earth, there would be no question that they were the ruins of a collapsed city.

Terrestrial Comparison

When a small section of the Western City Complex on Mars is compared to a bombed-out area of the German city of Berlin during World War II, the common destruction of a condensed section of a city is quite stunning (Fig. 6.32). Notice the partially collapsed architecture of free-standing walls and the hollowed-out buildings in both images. Could this be, yet again, another example of pars pro toto or "as above, so below?"

SEVEN

The Martian Atlantis II

District 1

Polygonal Mounds

My ANALYSIS OF the Atlantis Chaos region of Mars[1] continues by moving away from the main City Complex (labeled A in Fig. 7.1) to explore the area directly below it. Traveling over the smooth terrain I follow a path that takes me across two surface features that appear to be in vertical alignment with the City Complex. Those two include a circular mound (labeled B in Fig. 7.1) and a small crater of about the same size (labeled C in Fig. 7.1).

Further down, in a southeastern direction, there is a set of five polygonal formations that exhibit a high degree of geometry and symmetry. Although these polygonal mounds are highly eroded, they still maintain their geometric form. This group of mounds is so extraordinary in their shape and size they are recognizably out of place within the common topography that surrounds them. Considering the possibility that these formations may be the remains of another settlement related to the main City Complex, I have titled this southern area District 1 (labeled 1 in Fig. 7.1).

Unlike the tight cluster of relatively small walls and cubic foundations arranged in a grid-like design observed in the main City Complex, these formations are grouped together in a segregated area and are gigantic in size. Each formation is larger than the entire City Complex.

Taking a closer look at this group of geometrically shaped, polygonal mounds found just below the City Complex (labeled A in Fig. 7.1), I will begin

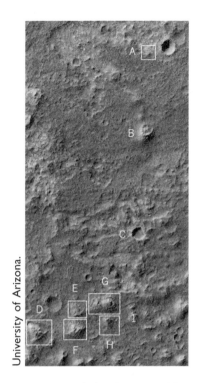

University of Arizona.

Fig. 7.1. The Martian Atlantis. Detail MRO
HiRISE ESP_019103_1460 (2010).
Notations by author.
A: The City Complex. B. Circular mound.
C. Crater.
1. District 1. Individual formations are
notated D–H.

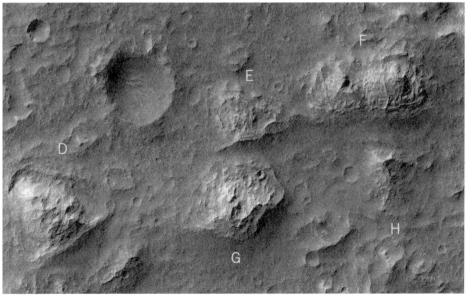

Fig. 7.2. District 1. Detail MRO HiRISE ESP_019103_1460 (2010). Individual
mounds are notated D, E, F, G, and H.

this portion of the study by examining each of these formations individually
(labeled D–H in Fig. 7.2).

The Soft Triangle

Starting with the formation labeled D in Fig. 7.2, which is located on the westernmost side of District 1 and sits on the edge of the image, we see the remains of a triangular-shaped foundation. This triangular formation is highly symmetrical and has soft, rounded corners (Fig. 7.3). If this formation was once produced as a fully functional archology, its triangular roof or dome covering appears to be missing, leaving its infrastructure fully exposed. Its highly eroded interior ceiling is now partially filled with silt and rubble. The center of the structure is elevated higher than its southwestern corner and its eastern side, which is smooth and possibly filled with silt. Notice the raised, linear edge of its triangular footprint, leaving the impression that the original supportive walls and roof were violently ripped off its foundation. Fig. 7.3 offers the proposed triangular geometry of the structures footprint.

Fig. 7.3. Soft Triangle (labeled D in Fig. 7.2).
Top: Detail MRO HiRISE ESP_019103_1460 (2010).
Bottom: Proposed geometry.

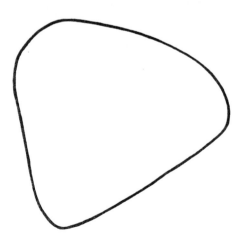

Terrestrial Comparison

When the Soft Triangle formation on Mars is compared to the modern architectural plans of the Riverside Center (Fig. 7.4), built by Harry Seidler in Queensland, Australia,[2] the conception of a triangular structure with rounded corners become quite plausible. The symmetry and triangular geometry of the two formations is exquisite in their soft form and shape. When the architectural plans of the Riverside Center are overlaid on top of the Soft Triangle formation on Mars their common triangular forms fit like a glove, round corners and all.

Considering the close grouping of the Soft Triangle with its adjoining formations, which are sitting right in front of it with various geometric shapes, it

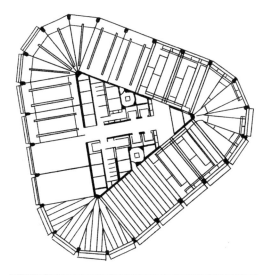

Fig. 7.4. Soft Triangle architectural overlay. Top: The Riverside Center. Queensland, Australia, 1986. Bottom: Soft Triangle on Mars with overlay of The Riverside Center, Architectural plans.

would be difficult to comprehend what type of natural erosional mechanism could have created this triangular form and not affect the geometric forms of its neighbors.

Acute Triangle

The next formation to be examined in this study of District 1 depicts the craggy remains of another large mound that takes on the form of a sharp triangle, labeled E in Fig. 7.2. Notice the acute angle of its triangular shape, which is one of the most basic shapes in geometry (Fig. 7.5). The interior surface is irregular and highly textured with a gouged-out, shallow impression just off its center line. Its southwestern corner has sharp edges and is filled with smooth, packed sand. The central land mass is rough and speckled with a fine, grid-like pattern that runs parallel with its northeastern side. Its western and northeastern sides are straight, while its southern edge is rippled, forming a repeated wave pattern that expands in size as it reaches its eastern point.

Despite its highly errored interior and slightly irregular footprint, its crumbling foundation still maintains its basic triangular shape. Even though

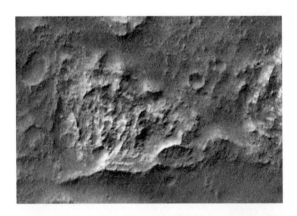

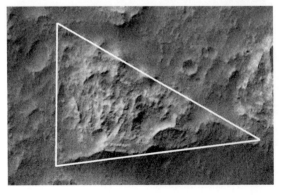

Fig. 7.5. Acute Triangle (labeled E in Fig. 7.2). Top: Detail MRO HiRISE ESP_019103_1460 (2010). Bottom: Proposed geometry. Detail MRO HiRISE ESP_019103_1460 (2010). Outline by the author.

its supportive walls and upper roof appear to have been worn away or blown off, its contours still fall in line with the geometry of an acute triangle. A study of its proposed geometry is provided in Fig. 7.5.

Terrestrial Comparison

This classic Acute Triangle structure, which is positioned at the northern center of District 1, can be compared to the triangular fortifications that were built in the towns of Liège, Namur, and Antwerp during the late 1800s (Fig. 7.6). Belgian architect Henri Alexis Brialmont designed a variety of polygonal and triangular forts that were constructed out of concrete with steel reinforcements that extended underground.[3] He believed that triangular forts had a strong defensive design and increased the ability of a turret to fire in all directions.[4]

The strength and viability of Brialmont's triangular forts would not be fully tested until the German invasion of Belgium during World Wars I and II. Unfortunately, every fort in their path was overrun and captured. Fig. 7.7

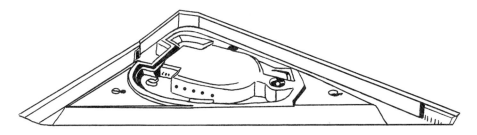

Fig. 7.6. Fort de Barchon. Liège, Belgium, circa 1887.
Designed by Henri Alexis Brialmont. Drawing by the author.

Photograph courtesy Marc Romanych.

Fig. 7.7. Fort de Barchon. Liège, Belgium, circa 1940.

provides an aerial photograph of one of the triangular forts after it was bombed by the German Army during World War II.

The Temple of Rectangular Blocks

The third structure in District 1 (labeled F in Fig. 7.2) takes on the form of a large rectangular platform supporting a pair of linear foundations that have a rectangular or blocklike shape (Fig. 7.8). Although the linear contour of this pair of block-shaped foundations is still visible the overall formation has been severely battered and highly degraded over time. Geologists will assert that it is not uncommon to find similar linear and box-shaped formations in nature; however, they will also agree that it is possible some of these linear contours and boxlike foundations may have been the result of something other than natural erosion.[5]

Taking a closer look at what I call the Temple of Rectangular Blocks (labeled F in Fig. 7.2), I began my study by highlighting the contours of these two rectangular foundations that were initially observed (Fig. 7.9). Once they were outlined, I not only noticed that one was a little larger than the other, but I also noticed they were not alone.

As I explored the entire formation, I found additional sets of the linear remains of foundations that can be seen throughout the formation. I noticed that the same block-shaped contours observed within the first two rectangular

Fig. 7.8. Temple of Rectangular Blocks (labeled F in Fig. 7.2).
Detail MRO HiRISE ESP_019103_1460 (2010).

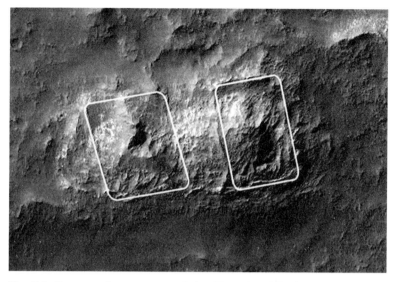

Fig. 7.9. Proposed geometry of the Temple of Rectangular Blocks
(labeled F in Fig. 7.2). Detail MRO HiRISE
ESP_019103_1460 (2010). Boxed outlined by the author.

foundations are repeated and superimposed, one on top of the other, throughout the formation. I suggest that this repeated pattern goes well beyond anything nature can produce. Fig. 7.10 shows the outline of seven linear foundations within the Temple of Rectangular Blocks.

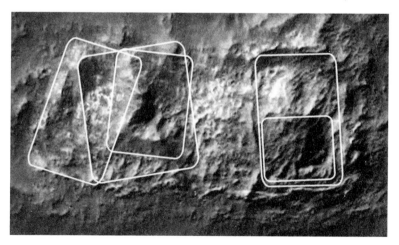

Fig. 7.10. Temple of Rectangular Blocks (labeled F in Fig. 7.2).
Detail MRO HiRISE ESP_019103_1460 (2010). Notice the seven rectangular mounds. Outlined with proposed repetitive geometry by the author.

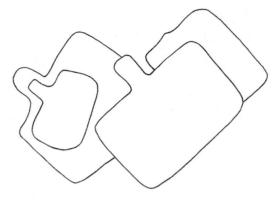

Fig. 7.11. Four superimposed pit houses. Area D at Mescal Wash, Tucson, Arizona, 2011. Drawing by the author after photograph by Statistical Research, Inc., Redlands, California.

Terrestrial Comparison

This repeated, superimposed pattern of rectangular foundations, as seen within the Temple of Rectangular Blocks on Mars, is very similar to the kind of designs found in fractal art and architecture. Fractals are the result of repeated geometric patterns that are "self-similar" and repeated in various sizes.[6] A great example of this type of fractal architecture can be seen within the construction of dwellings found in the deserts of North America. In the mid-1990s workmen discovered a group of foundations in southern Arizona during the construction of a new roadway. Hidden under layers of sand and dirt were the remains of numerous pit houses that were built by the Hohokam culture, who occupied the area between 750 to 950 CE.[7] During the extensive excavation of the area now known as the Mescal Wash Site, archaeologists uncovered a set of four superimposed pit houses that were built of various sizes, one on top of the other (Fig. 7.11). If I didn't know better, I would think I was looking at the remains of another Temple of Rectangular Blocks on Mars.

Arrowhead Mound

The fourth formation in District 1 to be examined in this study is a polygonal structure that has a very symmetrical arrowhead shape (labeled G in Fig. 7.2). Although its interior surface is rough and highly textured its outer contours have maintained its overall arrowhead shape (Fig. 7.12). The formation has seven sides with three pairs of adjacent sides of equal length. Its western side

Fig. 7.12. Arrowhead Mound (labeled G in Fig. 7.2). Top: Detail MRO HiRISE ESP_019103_1460 (2010). Bottom: Proposed geometry, outlined by the author.

forms a blade edge that conjoins to a point, while its eastern side forms a shoulder that extends down to the stem that forms a flat, vertical base. A study of its proposed geometry is provided in Fig. 7.12.

The interior topography of the Arrowhead Mound has an elevated center that is surrounded by blocky forms and linear walls that create a concentric pattern. Although much of its surface is cloaked by sand and debris, its clear arrowhead footprint invites our attention and demands an archaeological excavation.

Terrestrial Comparison

Searching through many examples of modern and avant-garde architectural designs I came across a promising candidate that can be compared to the Arrowhead Mound on Mars. I found a High-Tec, 3-D model of an arrowhead-shaped building that was created by Higyou at Alamy.com[8] (Fig. 7.13). Although the proposed model is streamlined and a little narrower in design, I think it can still be used as a conceptual equivalent. When looking at the geometry and the precise symmetry of the Arrowhead Mound on Mars, its

Fig. 7.13. Arrowhead Architecture.
Arrow-shaped building (3-D model). Drawing by the author.

form and shape is almost perfect. Traditionally, the sharp, abrasive design of an arrow is commonly used as a directional marker; however, in this context, it may have a more defensive function.

Triangular Pyramid with Bow-Tie Ramp

Just to the east of the Arrowhead Mound and directly below the Temple of Rectangular Blocks is the fifth structure of interest in this group of geometrically shaped structures (labeled H in Fig. 7.2). The structure consists of a triangle-shaped pyramidal mound with an attached bow-tie-shaped ramp (Fig. 7.14). The shape of the ramp also reminds me of those ancient stone-joint

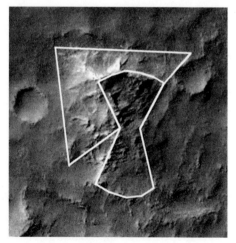

Fig. 7.14. Triangular Pyramid with Bow-Tie Ramp (labeled H in Fig. 7.2).
Left: Detail MRO HiRISE ESP_019103_1460 (2010).
Right: Proposed geometry, outlined by the author.

metal clamps used to bind masonry blocks. A study of the Triangular Pyramid with Bow-Tie Ramp with its proposed geometry is provided in Fig. 7.14.

Although this pyramidal structure has been extensively degraded over time, its overall triangular form can still be seen, along with the highly symmetrical geometry of its bow-tie-shaped ramp. If observed in isolation this odd formation may have been dismissed as a curiosity; however, due to its surrounding structures it merits a closer look.

Terrestrial Comparison

The imprint of the Triangular Pyramid with Bow-Tie Ramp may find its terrestrial counterpart in postmodern architecture. I think it can be compared to the folded repetitive design of origami architecture as seen in the innovative work of Polish American architect Daniel Libeskind (Fig. 7.15). Through his study of the ancient art form of origami Libeskind used its structural aspect of folding paper to solve architectural problems with dramatic results. He found that the angular design of origami opened a whole new realm of aesthetic form that creates spatial configurations, providing an effective tool for further explorations in the architectural design process.[9]

Most people were unaware of Daniel Libeskind's work before he won the 2003 competition for the reconstruction of the World Trade Center site in

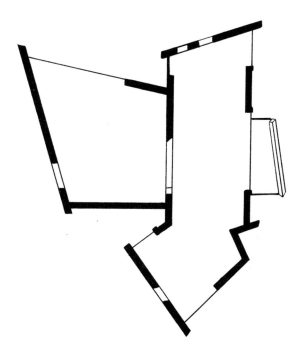

Fig. 7.15. Origami architecture. The Villa— Libeskind Signature Series by Daniel Libeskind. Datteln, Germany (2009). Drawing by the author.

Lower Manhattan. After visiting the site, he said, "Well, I was very moved when I walked around the site with millions of New Yorkers and trying to fathom and grasp what had happened on that site. And then, when I walked into the site itself, not only the indelible footprints of the towers were there, but something really amazing, which were the foundation walls."[10]

Is this what we are seeing on Mars? Are these geometric footprints of foundational walls on Mars, like the Triangular Pyramid with Bow-Tie Ramp, the remains of a once-grand city complex that suffered a similar fate?

Common Size and Measurements

On August 11, 2019, independent Mars researcher Michael J. Craig presented his interpretation of the size and shape of these polygonal features on his website Secret Mars, The Mars Archaeology Archive (Fig. 7.16). He identified four of the formations in District 1 and labeled them A–D. Craig's analysis not only confirmed my proposed geometric shape of the formations, but he also noticed a common size is shared between each of the structures, which he said suggests a deliberate relationship.[11]

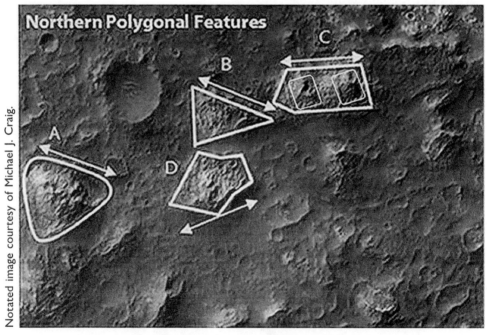

Fig. 7.16. District 1. Common size and measurements among polygonal features. A. Soft Triangle. B. Acute Triangle. C. Temple of Rectangular Blocks. D. Arrowhead Mound.

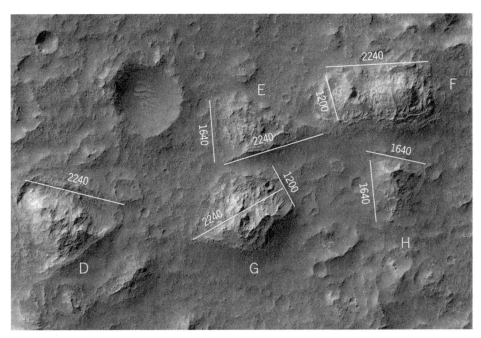

Fig. 7.17. District 1. Common size and measurements among polygonal features. D. Soft Triangle. E. Acute Triangle. F. Temple of Rectangular Blocks. G. Arrowhead Mound. H. Triangular Pyramid with Bow-Tie Ramp. Notated by James Miller and the author.

Looking at the original set of polygonal formations that make up District 1 (labeled D–H), I asked image specialist James S. Miller to perform his own analysis of each of the formations. Miller measured the individual size and dimensions of each of the formations and came back with some interesting results (Fig. 7.17).

Miller found that there are common measurements of 2,240, 1,640, and 1,200 meters within the length and widths of each of these five formations in District 1. Beginning with the Soft Triangle (labeled D in Fig. 7.17), it is oriented in a southwestern to northeastern orientation and measures approximately 2,240 meters wide on its three sides. The Acute Triangle (labeled E in Fig. 7.17) is aligned in a horizontal east to west orientation. It measures 2,240 meters along its northeastern and southern sides and measures 1,640 meters wide. The Temple of Rectangular Blocks (labeled F in Fig. 7.17) is also aligned along a horizontal west to east orientation. It measures 2,240 meters across its length and is approximately 1,200 meters wide. The Arrowhead Mound (labeled G in Fig. 7.17) is oriented on a northeastern to southwestern orientation along its point. It measures 2,240 meters in length, from its tip to its eastern base, which is also 1,200 meters wide. The origami architecture embedded within the Triangular Pyramid with Bow-Tie Ramp is aligned in a north to south orientation and measures 1,640 meters wide and 1,640 meters in length.[12]

Miller concluded that these common measurements provide evidence for a deliberate construction standard, the same type of uniformity that you would expect in the planning of a city or a residential development.[13] After comparing these large geometric structures to the much smaller compartmentalized walls seen within the main city area of the Martian Atlantis complex, Saunders speculated that they appear to be self-contained archologies that may have been constructed during a different time period.[14]

Taking a quick glance toward the south, I noticed that this district is not the only satellite settlement of large geometrically shaped structures to be found within the realm of the Atlantic Complex.

EIGHT

The Martian Atlantis III

District 2

Polygonal Mounds

NEXT, I'LL EXTEND MY ANALYSIS of the Atlantis Chaos region of Mars away from the southern edge of District 1 in the Mars Reconnaissance Orbiter HiRISE image[1] labeled 1 in Fig. 8.1 and survey the area directly below it. Descending southward to traverse a topography of small hillocks and mounds that bleed into a flat plan I see another set of enormous, polygonal formations that exhibit a high degree of geometry and symmetry.

Just as I observed with the cluster of five geometrically shaped formations within District 1, this second set of formations also appears to be archologies that are unusual and out of place within their surrounding environment. This group also includes a set of five formations that may be the remains of another settlement related to District 1 and the Twin Cities, located high above it to the north. Due to its proximity to the first set of formations, I have titled this second set District 2 (labeled 2 in Fig. 8.1 on page 148).

A cropped section of the image is provided in Fig. 8.2 on page 148; it shows the second set of polygonal mounds that I have identified here as District 2. In order to highlight their overall geometric shapes, I have outlined the proposed contours of each of these formations and labeled them I through M.

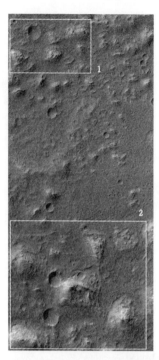

Fig. 8.1. The Martian Atlantis,
Districts 1 and 2. Detail MRO
HiRISE ESP_019103_1460 (2010).
Notated by the author.
1: District 1.
2: District 2.

Fig. 8.2. District 2. Polygonal mound formations with proposed geometry.
Detail MRO HiRISE ESP_019103_1460 (2010).
Outlined and notated I–M by the author.

Bulbous Octagon with Extended Platform

The first formation to be examined in District 2 is labeled I in Fig. 8.2. Its outer contours have an irregular octagonal shape with an extended platform on its eastern side (Fig. 8.3). The uneven border of its bulbous form consists of accumulated rubble and jagged angles, while its interior topography appears to be highly textured and extremely eroded in some areas.

The surface area just above the main body of the Bulbous Octagon with Extended Platform is defined by a sloping ridgeline containing small rippling dunes. Its lower, southern side is smoother, and much of its foundation appears to have either washed away or is covered in a fine sediment of sand and silt. Notice the highly textured stony interior of the Bulbous Octagon with Extended Platform is in extreme contrast to the smooth surface that surrounds it. A study of its proposed geometry is provided in Fig. 8.3.

Fig. 8.3. Bulbous Octagon with Extended Platform (labeled I in Fig. 8.2).
Top: Detail MRO HiRISE ESP_019103_1460 (2010).
Bottom: Proposed geometry, outlined.

 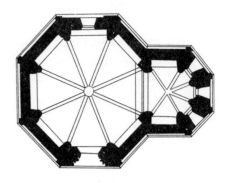

Fig. 8.4. Octagonal architecture.
Left: Bulbous Octagon with Extended Platform.
Detail MRO HiRISE ESP_019103_1460 (2010).
Right: Karner chapel. Oedenburg, Austria, circa thirteenth century.
Drawing by the author.

Terrestrial Comparison

Although the Bulbous Octagon with Extended Platform (labeled I in Fig. 8.2) is highly eroded with jagged contours along its north and western sides, its overall octagonal footprint is still traceable. Its classic shape resembles the octagonal shape often seen in the traditional designs of churches and chapels. Its architectural shape is seen as a religious templet and a symbol of rebirth and resurrection. Its form alludes to the shape of a water vessel, and its alcove is where baptismal altars were placed in many churches.[2]

A fine example of an octagonal church can be seen in the design of the Karner funerary chapel, which was built in Oedenburg, Austria, in the middle of the thirteenth century[3] (Fig. 8.4). Notice the octagonal shape of the main chapel and the extended alcove, which echoes the extended platform of the Martian formation.

Triangular Point

The second highly symmetrical formation in District 2, labeled J in Fig. 8.2, is a large triangular shape that tapers down to a sharp point (Fig. 8.5). The triangle is one of the most basic shapes in geometry and easily recognizable. Its northern end is flat and aligns horizontally with the north, while its western and eastern sides are of equal length and converge to a point toward the south. Notice the southern point appears to have broken off and slipped away from

Fig. 8.5. Triangular Point (labeled J in Fig. 8.2).
Left: Detail MRO HiRISE ESP_019103_1460 (2010).
Right: Proposed geometry, outlined.

the main foundation, giving the point an elongated appearance. A study of its proposed geometry is provided in Fig. 8.5.

The interior surface of the Triangular Point is rough and highly textured and has a thick northern ridge line that is followed by a smooth accumulation of sand and silt. This surface feature may be the result of a meteor impact seen just above its northeastern side. There are also the remains of two impact craters with a smooth basin floor in the upper center of the triangular formation. The triangular shape and the thick upper ridge give the formation the whimsical appearance of a slice of pizza with a thick crust and a slice of pepperoni.

Martian Comparison

The basic triangular shape of the Triangular Point (Fig. 8.6 on page 152) is very similar to the Acute Triangle that was seen in District 1 and discussed in the previous chapter. (See Fig. 7.5 on page 136.) Both formations have a distinct triangular shape and include the remains of an impact crater sitting just outside the corner of their shortest side. The two triangular formations are oriented in opposite directions. The Acute Triangle is oriented in a west to east direction, while the Triangular Point is set at a north to south alignment.

Fig. 8.6. Common triangular shape.
Left: Triangular Point (District 2). Detail MRO HiRISE ESP_019103_1460
(2010).
Right: Acute Triangle (District 1). Detail MRO HiRISE ESP_019103_1460
(2010), rotated.

Terrestrial Comparison

Just as we saw with the Acute Triangle, the contours and footprint of the Triangular Point (labeled J in Fig. 8.2) can be compared to the triangular military fortifications built by the famed French architect Henri Alexis Brialmont in the late 1800s. (See Fig. 7.6 on page 137.) Contemporary architects have also embraced the simplistic geometry of the triangle form (Fig. 8.7). The collaboration between the Dutch architecture firm Jo Coenen and a group called Archisquare in 2011 resulted in the design and construction of a triangular-shaped office building over the remains of an ancient Roman foundation in Parma, Italy.[4] When viewed within the context of triangular-shaped fortresses and architectural buildings, the idea that the Triangular Point formation on Mars could be a triangular-shaped arcology becomes much more plausible.

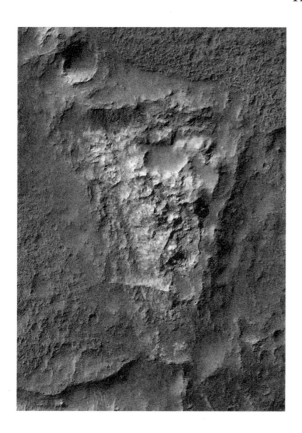

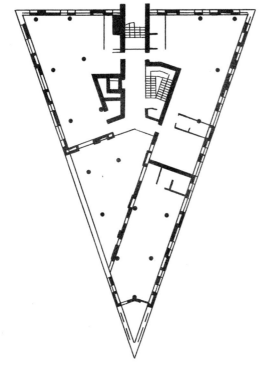

Fig. 8.7. Triangular Point (labeled J in Fig. 8.2). Top: Triangular Point. Detail MRO HiRISE ESP_019103_1460 (2010). Bottom: Triangular office building architectural plans. Drawing by the author after Jo Coenen and Archisquare.

Fig. 8.8. Chevron Shield (labeled K in Fig. 8.3).
Left: Detail MRO HiRISE ESP_019103_1460 (2010).
Right: Proposed geometry, outlined.

Chevron Shield

The third formation in this study of District 2, labeled K in Fig. 8.2, takes on the outward appearance of a polygonal formation with six equal sides that resembles a chevron-shaped shield (Fig. 8.8). Despite the collapse of its upper eastern corner, the overall shape is created by a combination of linear ridge lines and carved-out trenches that surround a highly erratic interior surface. There is also a triangular imprint with a mound of highly reflective material just above the northwestern trench. A study of its proposed geometry is provided in Fig. 8.8.

Martian Comparison

Traveling northward over to the Utopia region of Mars there is a very similar chevron-shaped formation that has been referred to as Envelope Island ever since it was discovered during the early Viking mission.[5] Fig. 8.9 is a Mars Odyssey THEMIS image of the formation. The image was taken in the winter of 2005, during the late morning, with a resolution of approximately nineteen meters per pixel.[6]

In comparing the shape of the two formations, the contours of the Chevron Shield appear to be gouged out and carved into the surface, while its interior is elevated and highly textured and pitted. In contrast, the construction of the Envelope Island formation appears to be a thick, highly elevated platform with

Fig. 8.9. Common chevron shape.
Top: Chevron Shield. Detail MRO HiRISE ESP_019103_1460 (2010).
Bottom: Envelope Island. Detail THEMIS V13700005 (2005).

an overall chevron shape. Its rough interior is speckled with smaller mounds of different sizes. Both formations have a common broken, indented lip along the top, giving it a slight V-shaped design.

Terrestrial Comparison

Within the vast lexicon of signs and symbols the wave-like shape of a chevron insignia is a pictographic representation of the movement of water. It is also seen as a symbol of stability and competence and has been used to distinguish

Map data © 2016 Google.

Fig. **8.10.** Common chevron shape.
Top: Chevron Shield. Detail MRO HiRISE ESP_019103_1460 (2010).
Bottom: Fort Douaumont, France.

military authority and rank.[7] As a military icon, the Chevron Shield on Mars (labeled K in Fig. 8.2) can be compared to a similar terrestrial formation known as Fort Douaumont. Constructed in 1885 the fort protected the French city of Verdun from assault.

Despite its long history of success, Fort Douaumont was partially destroyed in 1916 by the German army's bombardment of super-heavy howitzers during World War I.[8] Although nature has reclaimed much of the site over the past hundred years, the fort has been preserved, and its overall footprint can still be seen today (Fig. 8.10).

The common chevron design shared with its terrestrial companion is quite remarkable. You can see that although the ridge lines and trenches of the Martian formation conform to a chevron design, its wide foundation is a bit irregular and does not provide a complete footprint. Its eastern side appears to have been breached and partially distorted. I find it intriguing that the triangular imprint above the northwestern trench of the Chevron Shield has a mound of highly reflective material that echoes a similar formation observed in the Google Earth image of Fort Douaumont (Fig. 8.10).

Elongated Pentagon with Polygonal Depression

Moving down to the fourth and central formation observed within District 2 (labeled L in Fig. 8.2), it displays an elongated, irregular pentagonal shape that resembles a casket. Its western side has a raised mound form that descends into a polygonal-shaped depression (Fig. 8.11). The polygonal depression has seven sides and shows signs of bilateral symmetry. This feature may have been the result of a collapsed supportive structure that held up a dome feature or roof. Notice all the debris and rubble in the depression and the square-shaped mound in the upper center.

Terrestrial Comparison

The overall contours of the Elongated Pentagon with Polygonal Depression look a lot like the remains of a defensive structure produced by the Egyptians known as the Buhen fortress (Fig. 8.11 on page 158). Constructed during the Second Dynasty[9] the fortress has a five-sided polygonal shape. Unfortunately, it is situated on the west bank of the Nile and is now totally submerged underwater. Much like its Martian counterpart, the Buhen fortress has an elongated pentagonal shape and a large interior courtyard attached to an outer wall.

The Serpent Bell

The last formation in District 2 to be examined is a very large bell-shaped mound (labeled M in Fig. 8.2). The bell form is highly textured and is raised in its center, which has a slightly square-shaped area that is highly reflective (Fig. 8.12 on page 159). There is a thick depression that follows the contours of the bell shape forming an outer ridge line.

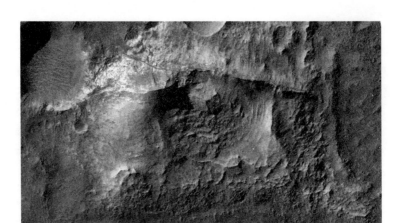

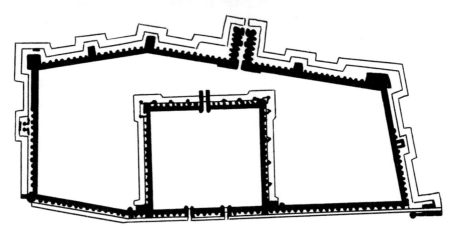

Fig. 8.11. Elongated Pentagon with Polygonal Depression (labeled L in Fig. 8.2).
Top: Detail MRO HiRISE ESP_019103_1460 (2010).
Middle: Proposed geometry, outlined.
Bottom: Buhen fortress, Egypt. Drawing by the author after Walter B. Emery.

Fig. 8.12. Serpent Bell (labeled M in Fig. 8.2).
Top: Detail MRO HiRISE ESP_019103_1460 (2010).
Bottom left: Proposed geometry, outlined.
Bottom right: Analytical drawing by the author.

A highly textured surface pattern that flows out across the surface can be seen at its base on the western side. The pattern takes on the imprint of a serpentine figure with an exposed neck and dragon head. Fig. 8.12 provides an analytical drawing of the formation's overall bell shape and its extending serpent figure.

If one zooms in on the serpent's head and neck section a rectangular-shaped eye socket with a round eye feature can be seen at its center (Fig. 8.13 on page 160). It has an open mouth with a beak-like upper and lower mandible.

Fig. 8.13 Serpent Bell, detail of neck and head.
Left: Detail MRO HiRISE ESP_019103_1460 (2010).
Right: Analytical drawing by the author.

The upper mandible has two triangular fangs, while the lower has one. The neck area is segmented with a scale and feathering pattern that extends up into the collar and head area. Unlike the other polygonal formations observed in Districts 1 and 2, this bell-shaped formation incorporates a pictographic representation of a serpent coming out of its western corner; therefore, I have called this formation the Serpent Bell.

Terrestrial Comparison

The design of the Serpent Bell formation is very similar to the Aztec style of glyphs that feature dome-shaped mountains or hills with an animal attached to its surface. A fitting example is found in the colorfully illustrated Codex Boturini* (Fig. 8.14). The codex shows a glyph depicting a serpent emerging from a bell-shaped mountain that is very similar to what I have found on Mars. Although the Aztec version shows a serpent emerging from the top of the mountain and the Martian version shows it emerging from the bottom, both

---

*The Codex Boturini is an Aztec book, which records the migration of the Aztec and Mexica, people from Aztlán. The unfinished work was produced before or just after the conquest of the Aztec Empire. The codex is currently located in the National Museum of Anthropology in Mexico City.

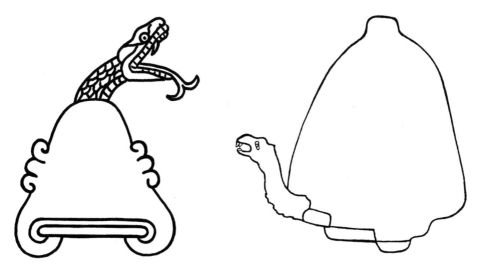

Fig. 8.14. Serpent Mountain comparison.
Left: Serpent Mountain, Aztec. Codex Boturini, page 10. Drawing by the author.
Right: Serpent Bell, Mars. Analytical drawing by the author.

share a common iconography that may provide further evidence of an ancient contact between these two worlds.

Common Size and Measurements

Just as he did with District 1, independent Mars researcher Michael J. Craig presented his interpretation of the size and shape of the polygonal formations of District 2 on his website, Secret Mars. In the article titled "The Mars Archaeology Archive," Craig identified four of the formations and labeled them E–I (Fig. 8.15). Craig's analysis not only confirmed much of my proposed geometric shapes of the formations, but he also noticed a common size is shared between each of the structures, which he suggests was a deliberate relationship.[10]

I approached image specialist James Miller and asked him to analyze the individual formations in District 2 and provide the measurements of the five polygonal formations that I have labeled I, J, K, L, and M (Fig. 8.16 on page 163). Miller found a template of common sizes and shared measurements between all five of the formations. He found a common measurement of 2,625 meters is shared within the width of each of these formations, and a common measurement of 2,900 meters shared along the length, showing evidence that there was

Fig. 8.15 District 2. Common size and measurements among polygonal features labeled E-I. Notated image courtesy of Michael J. Craig.

a deliberate, standardized measurement utilized in the construction of these formations.

Starting with the Bulbous Octagon (labeled I in Fig. 8.16) it is oriented in an east to west direction and measures approximately 2,625 meters wide and is 2,900 meters in length. The Triangular Point is oriented in a north to south direction (labeled J in Fig. 8.16) and measures 2,625 meters along its northern side and 2,900 meters in length from its northern side to its southern triangular point. The Chevron Shield (labeled K in Fig. 8.16) is oriented in a northwest to southeast orientation and measures approximately 2,625 meters across its width. The third formation identified as the Elongated Pentagon with Polygonal Depression (labeled L in Fig. 8.16) is oriented in a northwest to southeast orientation. It has a width of 2,625 meters and a length of 2,900 meters. The Serpent Bell is the largest of the formations in District 2 (labeled M in Fig. 8.16), and like the Triangular Point it is oriented in a north to south orientation. It measures 2,900 meters wide from the nape of the serpent's neck to the outer wall of the eastern side of bell form. From the top of the bell form to the bottom of the undulation posture of the serpent's body, it is a whopping 5,250 meters in length. Miller noted that the measurement of 5,250 meters is two times the standard 2,625-meter measurement found in its surrounding formations.[11]

Miller believes that aesthetically, when any set of architectural formations are constructed within the same size and dimensions, they will produce a unified appearance. Their common size creates a balanced rhythm and harmony within the landscape and projects a uniform aesthetic; therefore any unifying standard of size and measurements would be instrumental in the construc-

Fig. 8.16. Common size and measurements among polygonal features.
I. Bulbous Octagon. J. Triangular Point. K. Chevron Shield. L. Elongated Pentagon
with Polygonal Depression. M. The Serpent Bell. Notated by the author.

tion of any city or complex.[12] Finding this much uniformity and balance in nature is extremely unlikely. When it comes to finding any type of architecture signature within any given landscape, the presentation of geometry would be inevitable in its organization.

I find it quite compelling that all of these polygonal structures that occupy the Martian Atlantis Complex, including the boxlike structures that make up the Twin Cities and the massive archologies observed in both Districts 1 and 2, are all found in such a small, condensed area. Their mere presences leave us with only one of two options. The first option suggests that this geological anomaly has an uncanny propensity to produce highly symmetrical landforms that are not only geometric in form but also exhibit a common set of dimensions. Our second option proposes the idea that someone with a vast knowledge of architectural design had a hand in their creation.

NINE

Parrotopia I

Wilmer Faust

ON MARCH 7, 2002, independent researcher Wilmer Faust presented an odd hillock formation captured in a Mars Global Surveyor image to the members of two research groups. It was sent to The Cydonia Institute and to The Anomaly Hunters, which was founded in 2002 by James S. Miller. Similar to my group, Miller's was organized as a cooperative for the study of anomalous formations observed in Mars image data. His website was shut down in 2009 and resurrected in the summer of 2021 as an online Facebook group.

The Mars Global Surveyor image that Faust presented to the two groups is shown in Fig. 9.2. It was taken in the spring, during the early evening, with a resolution of 2.77 pixels per meter.[1] Faust directed our attention to the compartmentalized structural features throughout the area's topography as well as a formation of entirely different geometry that was suggestive of a gigantic profile of a bird. Faust identified a mound-shaped body with a head that included an eye and beak. He noticed the body featured a leg and foot and an extended wing with feathers. Faust thought the shaping of the hillside wing was nearly a perfect rendering of the conformation of a parrot's wing feathers. He also acknowledged that although the approximate shape of a parrot's midsection might be due to chance erosional forces, its detail was simply too uncannily realistic. He also made note of very fine feathers above the eye that accurately correspond to the typical boundary of an avian eye patch.[2]

After seeing the image, I immediately saw the parrot formation and was so impressed with its detail that I began working on an article titled "Parrotopia."

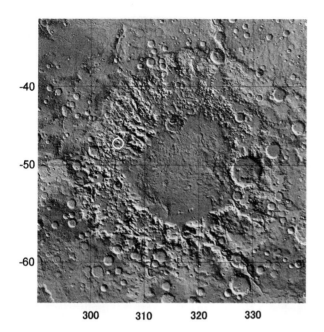

-40

-50

-60

300   310   320   330

Fig. 9.1. Argyre Basin, NASA MOLA Data map. Notated with the approximate location of the avian-shaped formation. Courtesy NASA/JPL/Malin Space Science Systems/The Cydonia Institute. Annotated by the author.

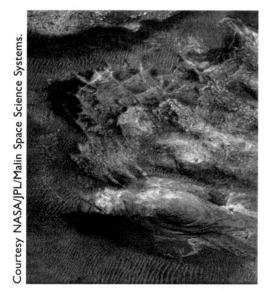

Courtesy NASA/JPL/Malin Space Science Systems.

Fig. 9.2. Parrot Geoglyph. Detail MGS M14-02185 (2000).

I posted an early draft of the article on the Project Teardrop* section of The Cydonia Institute website that November.

---

*Project Teardrop was a subsidiary section of The Cydonia Institute that focused on anomalous formations outside of the Cydonia region of Mars. Its title was inspired by the response of a NASA scientist who said after seeing the new MGS images of Mars, it brought tears to his eyes. The site was retired in 2004.

A Second MGS Image

Another MGS image (Fig. 9.3) was acquired in December of 2005 and had a "product creation time" of 06-20-2006 but was not publicly released until August 22, 2009. I was really perplexed as to the long lag time in its release. So I decided to inquire with F. Kuehnel at NASA's PDS image archive in September 2009. I asked him if the "product creation time" of 06-20-2006 was the official release date was for MOC image S13/01480. Kuehnel explained that the "product creation time" of 06-20-2006 was not the official product release date but could not comment on the later release date from August 2009. As a result, the conflict between the image being acquired in December 2005 and not released until August 2009 is still not resolved.

The second image (Fig. 9.3) of the Parrot Geoglyph is fantastic. It provides a clear and more complete image of the avian formation with an exceptional resolution of 1.43 meters per pixel. It shows the entire avian feature, including its head, body, and the extent of the tail feature. It shows the head, eye, beak, jaw, and tongue. The oval-shaped body has a foot with claws and an upper wing and extending tail feathers. The image is so good it leaves little doubt that this is an actual construct.

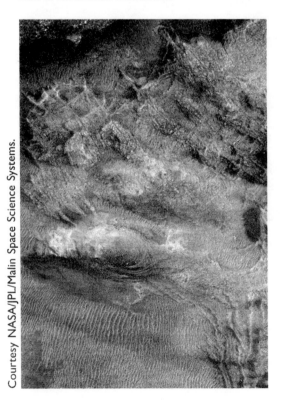

Courtesy NASA/JPL/Malin Space Science Systems.

Fig. 9.3. Parrot Geoglyph.
Detail MGS S13-01480 (2005).

MRO

Another image of the Parrot Geoglyph (Fig. 9.4) was acquired in the summer of 2013, during the early evening, with a resolution of 5.0 meters per pixel. This CTX image confirms all the physical features of the Parrot Geoglyph observed within the earlier MGS images. The details of the head, including the eye, beak, jaw, crest, and tongue, are all perceptible. The right leg structure is visible, and the upper leg and the contours of the foot and toes are still present. This MRO HiRISE CTX image also provides an extended view of the entire tail feature.

Although there are considerable differences in the telemetry, sun angle, and resolution in these three images, the basic physical feature of the avian formation remains persistent in each image.

Soon after the release of the MRO HiRISE CTX image of the Parrot, independent Mars researcher R. DeRosa filed a targeting request for additional images of the formation at the Mars Reconnaissance Orbiter HiWish site on March 9, 2014.[3] His petition was simply titled "Region within Nereidum Montes" and once filed, it would take another six years before the HiWish team granted his targeting request. The long-awaited MRO HiRISE image was released on January 14, 2021. Amazingly the image was titled "Parrotopia" at the HiWish site.[4]

This new image (Fig. 9.5) was acquired in the winter during the early afternoon with a resolution of 50.4 cm per pixel and provides the highest resolution to date. It confirms all the anatomical features observed in the earlier images with exquisite detail. The image also provides evidence for a second foot with a set of splayed toes at the top of the body.

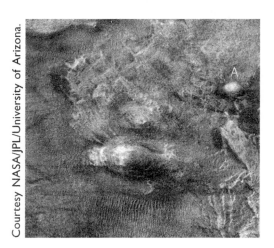

Courtesy NASA/JPL/University of Arizona.

Fig. 9.4. Parrot formation. Detail MRO HiRISE CTX D15_033142_1288_XI_51S054W (2013).

Argyre Basin

The Argyre Basin is located within the Argyre Planitia region of Mars. It encompasses a large-impact crater located in the southern hemisphere of the planet, below and to the southeast of Valles Marineras between -42° and -57°S and 326°E and 306°W[5] (Fig. 9.1). The impact basin is approximately 1,100 kilometers in diameter and is believed to have been created in the earliest period of Mars's geologic history about four billion years ago.[6] A rapid melting of the south polar ice cap is believed to be responsible for the basin to become water-filled during the Noachian Period.[7] In Greek mythology it was an island situated in the Indian Ocean and was said to be made of silver.[8]

Geological Analysis

A geologist, Michael Dale, and a geomorphologist, William Saunders, have examined MGS and MRO images of the Parrot Geoglyph[9] and found it to be composed of eight segments (Fig. 9.6). The segments include an extended right wing (1), beak (2), face (3), neck (4), body (5), lower leg and foot (6) tail feathers (7), and second leg/foot (8). These segments are differentiated by height, color, patterning, contour, and lithology. The central mound that forms the body and tail (labeled 5 in Fig. 9.6) is sedimentary in appearance. Since the height of the mound is roughly 175 meters[10] and sand dunes on Mars are typically only 10 to 25 meters in height,[11] an eolian depositional feature can likely be ruled out. The southwestern quadrant of the Argyre crater is suggested to have numerous glacial features including eskers.[12]

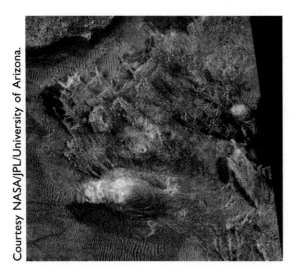

Courtesy NASA/JPL/University of Arizona.

Fig. 9.5. Parrot Geoglyph. Detail MRO HiRISE ESP_020794_1860 (2021).

Saunders and Dale contend that although the avian feature is in the northwest quadrant, the subglacial deposition in the form of a drumlin or esker that has undergone lithification would be the most likely candidates for the formation of the avian feature's body. The layered or stratified appearance that gives the visual impression of bird feathers is similar to what could be formed through wind or water action with the feature undergoing postdepositional erosion. The extended right wing (labeled 1 in Fig. 9.6) is highly textured in appearance with longitudinal and shorter perpendicular and slightly angular fractures. This is obviously different lithology than the body, and its extensive fracturing is likely due to rapid cooling. What appears to be a block fault separates the beak from the face forming the mouth (labeled 2 in Fig. 9.6). Postfaulting depositional material is the most probable natural explanation for the tongue identified within the fault cavity. The composition of the beak (labeled 2 in Fig. 9.6) could be composed of the same material as the body (labeled 5 in Fig. 9.6), having been separated by the removal of material from the face area.[13]

The avian-shaped mound is truncated at the neck, leaving the portion of the structure between the neck and the beak structurally lower and forming

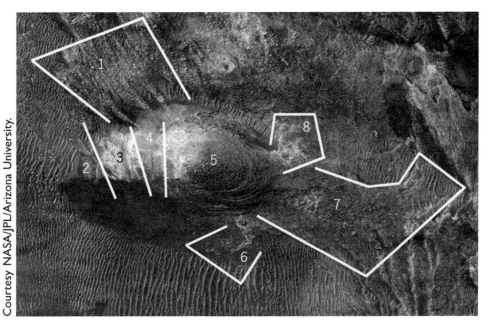

Courtesy NASA/JPL/Arizona University.

Fig. 9.6. Eight Segments of the Parrot Geoglyph. Detailed crop of MRO HiRISE ESP_020794_1860 (2021). Notation and line annotations by the author.
1) Extended right wing 2) Beak 3) Face 4) Neck 5) Body 6) First leg/feet
7) Tail feathers 8) Second leg/feet.

the face (labeled 3 in Fig. 9.6) by the exposure of an older underlying material, possibly from a lava flow. Interestingly, there is no wind-deposited material covering the face; however, wind action may be responsible for a darker material that appears to have been deposited up against the truncated body forming the hood or neck (labeled 4 in Fig. 9.6). The truncated and irregular edge at the juncture of the face and neck raises the question whether further erosion over the face occurred after this material was laid down.[14]

Possibly the most interesting feature in the aspects of its structure and exposure is the right leg and foot (labeled 6 in Fig. 9.6). The lighter color and structural level are similar to the face, indicating it likely consists of the same lithology. Interestingly, it remains exposed and has no windblown material obscuring it. The angular nature of the leg and toes would most conceivably be due to multidirectional faulting occurring prior to the deposition of the mound that forms the body. A possible left foot with splayed toes (labeled 8 in Fig. 9.6) also has a light color, again indicating it likely consists of the same lithology as the face. The tail section flows toward the east[15] (labeled 7 in Fig. 9.6) and conflates into the surface.

Their investigation concluded that it is apparent that the necessary geological and geomorphological processes to produce the avian-shaped feature took place in the Argyre Basin, however the procession and precise distribution of the geological events needed to produce all the avian features in their present form and proportion is highly unlikely. One geological process would destroy the previous and so on. The formation appears to be the result of a composite structure of unrelated geological materials that have been transformed into a sculptural relief that express the prominent features of an avian creature when observed from above.

Veterinarian Analysis

I contacted four veterinarians who agreed to provide an independent and impartial evaluation of the proposed avian formation observed within the Argyre Basin area of Mars. The four veterinarians include Dr. Amelia J. Cole in Virginia, Dr. Joseph Friedlander in New Jersey, Dr. Susan Orosz in Ohio, and Dr. Erica Mollica in New York. They were each provided with access to JPEG images of the complete avian formation as obtained in NASA MGS and MRO images and an analytical drawing produced by the author highlighting its features. The group found the avian formation to be a sculpted formation containing twenty-two points of anatomical correctness.[16] They also

Image courtesy of Phil Hart.

Fig. 9.7. King Parrot
(*Alisterus scapularis*).

acknowledge that although the earliest fossil of a parrotlike bird dates to the late Cretaceous about 70 million years ago,[17] it is reasonable to suggest that of the 353 known species of parrots, the avian formation on Mars shares its anatomical template with the terrestrial King Parrot[18] (Fig. 9.7).

The anatomical features in the image observed by the veterinarians include an oval-shaped mound that conforms to the shape and size of a bird's body resting on a folded right wing (Fig. 9.8 on page 172, point P). The avian formation is presented on its back with its abdomen exposed (Fig. 9.10, point Q), while its left-wing form (Fig. 9.8, point F) extends high above its body. On the left side of the mound-shaped body is a composite of structural elements that resemble a bird's head (Fig. 9.8, point S). The head includes the profiled view of an eye formation (Fig. 9.8, point D) and a large, parted beak (Fig. 9.8, point A) with a lower jaw (Fig. 9.8, point T). There is also evidence of a tongue (Fig. 9.8, point U). The beak has a feather-like protuberance, referred to as a crest by Dr. Cole and Dr. Orosz, that extends from the beak and projects out from the head (Fig. 9.8, point C). The modeling of the head is complex in its expression of texture and shading. The foreshortened orientation of the eye is remarkable in its proportion to the sightline expected within a profiled perspective. The plasticity of the beak appears hard and mantled, while the overall head and neck has a soft cauliflower look. Additional elements form an extended right leg (Fig. 9.8, points L and M) and clawed foot with exceptional adherence to muscular definition at the knee joint (Fig. 9.8, points N and O). The sculptural process of the leg appears to be fashioned in low relief and in effect has allowed sediment to cover portions of the detail. Attached to the body is an extended left wing along the back (Fig. 9.8, point F) with exposed feather

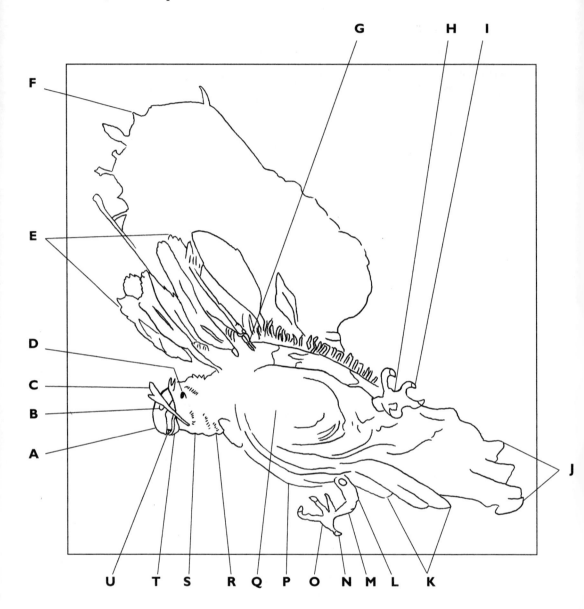

Fig. 9.8. Parrot Geoglyph. MRO HiRISE
ESP_020794_1860 (2021).
A) Beak B) Cere C) Crest D) Eye
E) Primary flight feathers F) Expanded wing
G) Feather shaft H) Right foot and toes I) Claw
J) Tail feathers K) Upper tail feathers L) Tibia M) Tarsus joint
N) Claw O) Left foot and toes P) Folded right wing
Q) Abdomen R) Neck S) Head T) Jaw U) Tongue.
Analytical drawing by the author with notations by A. J. Cole, DVM.

shafts (Fig. 9.8, point G). Behind the right foot are a set of upper tail feathers (Fig. 9.8, points K) that are again sculpted in low relief. The main tail extends from the body ending with splayed tips[19] (Fig. 9.8, point J). A second foot with splayed toes is positioned along the northeastern side of the main body, at the rump[20] (Fig. 9.8, points H and I).

Terrestrial Comparison

Around 400 BCE the Hopewell Indians of Ohio produced a small, hammered copper plaque[21] that has a very similar avian form that we see with the Parrot Geoglyph on Mars. In reviewing this Hopewell plaque (Fig. 9.9), avian specialist Dr. Orosz identified the form as representing an indigenous parrot. She acknowledges the overall profiled posture of the Hopewell parrot with its extended wing motif is reminiscent of the design expressed within the avian feature on Mars. She also notes the shape of the parrot's head and beak shares a common form with the avian feature on Mars, while the shape of the clawed foot, the round belly, and the stylized tail feathers are also analogous.[22]

There are many avian-shaped geoglyphs in North America. One of the most famous is a five-thousand-year-old formation in the town of Eatonton,

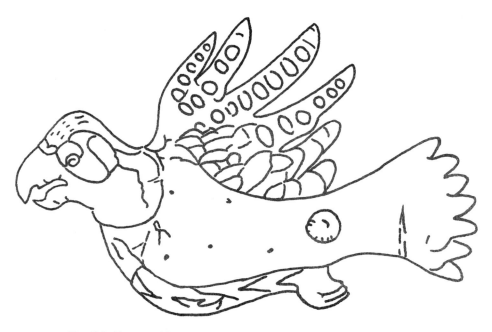

Fig. 9.9. Parrot. Hammered copper plaque (Hopewell Indians).
Drawing by the author after a photograph by Tom Engberg.

Fig. 9.10. Eagle Effigy Mound: Eatonton, Georgia. Image source: *National Geographic*, 142(6), page 784. Drawing by the author.

Georgia (Fig. 9.10). The avian formation is produced by a bed of white quartz stones, eight feet high. It forms a silhouette of an eagle hovering within a circular mound. The apex of the mound forms the eagle's abdomen, which creates a similar elevation as seen in the mound-shaped abdomen of the avian formation on Mars. The body of the eagle effigy measures more than 100 feet from head to tail and has a wingspan of more than 120 feet[23] (Fig. 9.10). The overall shape of the eagle effigy is symmetrical in design, featuring a set of outstretched wings, tail feathers, and a head that faces eastward. As seen in the illustration, its contours project only the simplest form of a bird without providing additional details.

Another example of an avian geoglyph is etched on a hillside in the Peruvian Andes, not far from the famous Nazca Lines.[24] The Peruvian pictograph is formed by a set of conjoined lines that create the impression of a standing bird (Fig. 9.11). Although the awkward shape of the Peruvian pictograph is not proportioned or anatomically correct, the overwhelming consensus is that it indeed represents the generic form of a small bird.

Accepting the consensus that these simple mound and hillside renderings are accepted as intentional works of art by the limits of aerial observations, it would be reasonable to suggest that the formal organization expressed within the Parrot Geoglyph on Mars conflicts with the randomness of mere chance. There are no terrestrial geoglyphs that induce such a visual impression that approaches the fine modeling of the sculptural relief as seen within the Parrot Geoglyph at Argyre Basin.

This Parrot Geoglyph has been documented in four images in this analysis that were taken by two different NASA spacecraft, the Mars Global Surveyor and Mars Reconnaissance Orbiter. They were taken at four different times and seasons and over a twenty-year period. With respect to the modeling of its anatomical features, they are accurately depicted and remain clearly visible. All its features appear to have permanence and are not the result of a transient phenomenon or an illusionary projection. Four veterinarians have found the formation to be exceptional in its physical appearance and anatomical completeness. While there are known geological mechanisms that can create the anatomical accuracies presented in this formation, the natural creation of a formation with an astounding twenty-two points of anatomical correctness seems to go well beyond the probability of chance. Therefore, I contend that the unique avian components of this Parrot Geoglyph are real and exhibit a level of consistency that is highly suspected of having artificial origins.

Since its original discovery by Wil Faust back in 2002, the Parrot Geoglyph has maintained a public and scientific interest. It has been the subject of two science papers and various books and magazine articles. NASA lead scientist of the Mars Orbiter Camera, Michael C. Malin, featured it on the NASA/JPL website's Image of the Day on August 22, 2005[25] and it was also mentioned on the front page of the *Wall Street Journal* in 2012.[26] The Parrot Geoglyph has been showcased on the History Channel's *Ancient Aliens* and *The Proof is Out There* programs and, most importantly, as a "tip of the hat" to its discoverer, NASA has titled the area in which the formation is located on Mars "Parrotopia."

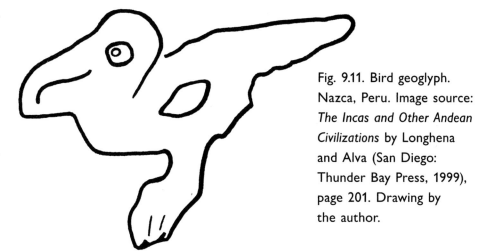

Fig. 9.11. Bird geoglyph. Nazca, Peru. Image source: *The Incas and Other Andean Civilizations* by Longhena and Alva (San Diego: Thunder Bay Press, 1999), page 201. Drawing by the author.

TEN

Parrotopia II
The Wing Complex

Compartmentalized Infrastructure

ZOOMING OUT FROM THE MAIN BODY of the Parrot Geoglyph this portion of the study will focus on the network of compartmentalized infrastructure that was initially observed by Wil Faust within the upper wing section of the Mars Global Surveyor image in Fig. 10.1. This MGS image was taken in the spring of 2000 during the early evening with a resolution of 2.77 pixels per meter.

Courtesy Wil Faust.

Fig. 10.1. Compartmentalized wing.
Detail MGS M14-02185 (2000).

Faust noted the chaotic complexity of the topography and believed it to be the remains of a harbor city. He suggested that the western side of the wing section bears a strong resemblance to jetties and breakwaters and saw harbor-related structures within its interior that are partially buried by various materials. He was convinced that the wing complex was the result of ancient intelligence and when the topography of the Argyre Basin is sufficiently understood by planetary scientists they should agree with him. He predicted they will conclude that high stands of water in this basin formed one or more equipotential elevations, marked by beach terraces will coincide with the mean elevation of the Parrot Geoglyph, providing a perfect place for a city with water frontage.[1]

Utilizing the higher-resolution MGS image S13-011480, which was acquired in the winter of 2009 with a resolution of 1.43 meters per pixel, one can observe some of the most interesting and anomalous features within the Wing Complex (Fig. 10.2). When MGS image S13-011480 is compared to the earlier 2000 M14-02185 image of the upper wing section, the structures seem to be compressed, while the M14-02185 image appears elongated and slightly askew. The slight difference in perspective is due to the low emission angle of the M14-02185 image, which is recorded at 0.29°,[2] while the S13-011480 image is recoded at 9.75°.

Image analyst James S. Miller explains that emission angle (EMA) is the angle between the spacecraft and a vector drawn perpendicular to the planet's surface. NASA often takes images by convenience of the positioning of the orbiters, which doesn't always provide the optimal angle or lighting. Often an image that has a definitive appearance can be lost when the image is taken at different times of year and from a different angle. This provides researchers with information that can support their observations or defeat them. These two MGS images are a prime example of this different imaging.[3]

Port Jetty

I will start with my examination of the Wing Complex by taking a closer look at the shoreline of the western and northwestern edge of the upper wing form. I direct your attention to the internal grid and projecting barb-like structures that bear a strong resemblance to jetties and breakwaters of a once functional port harbor. Fig. 10.3 shows the detail of the Port Jetty.

In 2021 the Mars Reconnaissance Orbiter HiRISE camera captured its first image of the Parrot Geoglyph, which included the entire wing section.

Fig. 10.2. Wing Complex, labeled A–G.
Detail MGS S13-01480 (2005).

Fig. 10.3. Port Jetty (labeled A in Fig. 10.2).
Detail MGS S13-01480 (2005).

University of Arizona.

Fig. 10.4. Port Jetty
(labeled A in Fig. 10.2).
Detail MRO HiRISE
ESP_067824_1320 (2021).

The image in Fig. 10.4 was taken in the winter during the early morning with a resolution of 50.4 cm per pixel. This new MRO HiRISE image confirmed all the internal grid and projecting barb-like structures observed in the Port Jetty section of the complex.

Terrestrial Comparison

The Port Jetty on the northwestern side of the Wing Complex is around 150 yards square, and it can be compared to the jagged and spiked design of a terrestrial docking port found in Canada. The port known as the Canadian Forces Base (CFB) is located at the southern tip of Vancouver Island (Fig. 10.5 on page 180). It is home to Canada's Pacific Coast naval base and encompasses 12,000 acres and 1,500 buildings.[4] While the naval base in Canada is much larger, the comparison is still valid.

The Water Jaguar

Moving away from the beams and jetties projecting out of the western edge of the wing form, I direct your attention to the partial image of a feline head, labeled B in Fig. 10.2. The beams of the jagged jetty act as patches of fur and the lightning spiked feathers of a decorative crest. The front of the feline's face is pressed up against a long, dark crack in the surface that runs up in a northern direction. It has a dark arched cavern that forms an eye. There is evidence of a muzzle that has a crescent-shaped nose ornamented with an extended bar. It has a slightly parted mouth with a small tooth, and a flailing tongue that extends down to a fuzzy chin.

Fig. 10.5. Port Jetty comparison.
Top: Detail MGS S13-01480 (2005).
Bottom: Canadian Forces Base (CFB). Vancouver Island, Canada.

The crack in the surface slices the face in half. In an effort to complete the image, I duplicated the partial facial formation, which revealed a frontal view of a jaguar with a spiky headdress (Fig. 10.6). Due to his close positioning within the Port Jetty, I have titled this feline geoglyph the Water Jaguar. The portrait of the Water Jaguar is shown to be consistent and is reproduced in the higher-resolution MRO HiRISE version (Fig. 10.7). An analytical drawing is provided in Fig. 10.8 on page 183 with a gray wash clarifying its face.

Fig. 10.6. The Water Jaguar (labeled B in Fig. 10.2.).
Detail MGS S13-01480 (2005), duplicated.

Fig. 10.7. Water Jaguar (labeled B in Fig. 10.2.).
Detail MRO HiRISE ESP_067824_1320 (2021), duplicated.

Terrestrial Comparison

The Water Jaguar, with its headdress of feathered lightning rods, reminds me of the Zapotec rain god known as Cocijo. He is a zoomorphic creature and is at times depicted as a jaguar with a flailing tongue and wears a radiant headdress with a water emblem on his forehead that archaeologists refer to as the C-glyph.[5] A great example of the Zapotec rain god in his jaguar aspect is in the Frissell Museum of Zapotec Art in San Poplo Villa de Mitle, Mexico (Fig. 10.9). Notice the large radiating headdress and the water emblem on his forehead.

The Eye

Moving further inland away from the Water Jaguar, I direct your attention to the human-shaped eye formation (labeled C in Fig. 10.2). Here is a close-up detail of the port eye formation (Fig. 10.10 on page 184). Notice the slim almond-shaped eye form and the decorative linear features that surround it. The eye form has an iris and a hooded lid form. The eye formation also has a long narrow berm that juts down from the lower edge of the eye that ends with a circular moat. There is also a linear, fish-fin-shaped ornamentation attached to the outer left side of the eye, resembling Egyptian mascara. The complete eye formation is also seen in the MRO HiRISE image (Fig. 10.11 on page 184).

Terrestrial Comparison

The human eye has been used in many artistic forms, such as the classic CBS "Eye" logo produced by graphic designer William Golden (Fig. 10.12a on page 184). Produced in 1951, the eye image is simple and direct. It has also been used in the design of anthropomorphic architecture. Spanish architect Luis de Garrido built an eye-shaped building in New Zealand (Fig. 10.12b on page 184). The eye-shaped building features a central glass-domed structure within adjoining supportive pavements that complete the overall eye shape, which was inspired by the design of the Egyptian Eye of Ra.[6]

The Rain God (The Eye of the Storm)

After finding this gigantic carving of an eye sitting within this port city right next to the Water Jaguar, I began pondering the thought that maybe this "eye"

Fig. 10.8. Water Jaguar. Analytical drawing and gray wash by the author.

Fig. 10.9. Jaguar Effigy (Cocijo). Frissell Museum of Zapotec Art, San Poplo Villa de Mitĺa, Mexico. Drawing by the author.

Fig. 10.10. The Eye
(labeled C in Fig. 10.2).
Detail MGS S13-01480
(2005).

Fig. 10.11. The Eye
(labeled C in Fig. 10.2).
Detail MRO HiRISE
ESP_067824_1320 (2021).

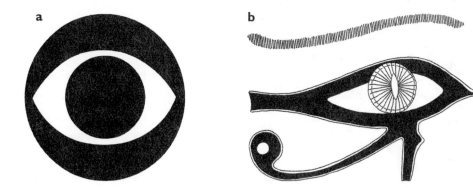

Fig. 10.12. Eye designs.
a. CBS logo (1951).
b. Eye of Ra, building. Luis de Garrido. New Zealand, 2011.
Drawings by the author.

is part of a much larger figurative formation. I noticed the eye sits within a half-moon section of the complex and has a partial nose and mouth below it. It also has a dark crack cut into the surface, much like the one I noticed along the jaguar's face. Using it as a guide, I duplicated the facial features and revealed a portrait of a larger figure (Fig. 10.13). Notice the head with an elaborate spiky headdress and truncated bust. The duplicated image has a face with two eyes, a nose with a cross bar, and a mouth with long hanging fangs. The highly reflective, crystalline torso has sharp shoulder pads and a diamond-shaped breast plate. Because its discovery began with the eye, I refer to this formation as The Rain God (The Eye of the Storm). All the facial features and the elaborate headdress are confirmed in the MRO HiRISE image (Fig. 10.14).

Fig. 10.13. The Rain God (Eye of the Storm). Detail MGS S13-01480 (2005), duplicated.

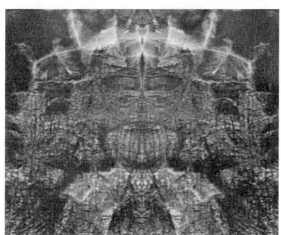

Fig. 10.14. The Rain God (Eye of the Storm). Detail MRO HiRISE ESP_067824_1320 (2021), duplicated.

Terrestrial Comparison

After I took another look at the facial features observed within The Rain God on Mars, I realized it looks remarkably like the Aztec god of lightning and rain known as Tláloc and as Chaac to the Maya. The Rain God is typically seen either holding a lightning bolt or wielding an axe. He is often depicted with goggled eyes, a small nose with a horizontal nose ornament, and feather-like electrodes emitting from his headdress. He also has odd mouth features such as jaguar teeth and long beard-like fangs hanging down.[7]

The National Museum of Anthropology in Mexico City has a ceramic brazier of the rain god that comes from the Gulf Coast region of Mexico that is quite comparable. It has the same facial features and spiked headdress that are seen within the portrait of The Rain God on Mars (Fig. 10.15).

So, one of the first questions one might ask is: How is the Rain God connected to the parrot, and how does his portrait end up in this Wing Complex? An answer to this odd relationship may be found in both Maya and Mixtec artifacts.

Fig. 10.15. Tláloc, Rain God (Veracruz). Detail of ceramic brazier. Image source the National Museum of Anthropology, Mexico City. Drawing by the author.

Fig. 10.16. Rain God as parrot. Detail of the Codex Zouche-Nuttall, page 76. Drawing by the author.

The first example takes us to a scene illustrated in the Codex Zouche-Nuttall.* The Mixtec book depicts the Rain God in the guise of a cute little parrot sitting on the Mountain of Sustenance (Fig. 10.16). He sits on a mountain that is filled with cacao and maize. Notice the glyph that is attached to the string of dots on the left. This is a Rain God glyph, which recognizes the parrot in this scene as a "lighting beast,"[8] which is an alternate personification of the Rain God.

According to creation stories the Rain God (seen here as a parrot) attempts to break open this sacred mountain by using its powerful thunderbolts but is unsuccessful. After his failed attempt the woodpecker swoops down and pecks the mountain with its powerful beak and splits its mantle open, releasing the sacred food.[9]

A more impactful depiction of the Rain God in his parrot aspect can be seen on a colorful Maya vase showing him as a monstrous bird of thunder and rain[10] (Fig. 10.17). Notice the parrot has the typical goggle-shaped eye and

---

*The Codex Zouche-Nuttall is a Mixtec, accordion-folded book that records the genealogies, alliances, and conquests of several eleventh- and twelfth-century rulers of a small Mixtec city-state located in highland of Oaxaca, Mexico. It now resides in the British Museum.

Fig. 10.17. The Rain God as a parrot. Detail of vase.
Drawing by the author after photograph by Justin Kerr K6809.

there is a Tláloc skull glyph with a giant headdress on his beak of sharp, jagged teeth. There is also a star glyph above his raised foot, and another is attached to his wing.

After reviewing these Pre-Columbian artifacts I'm confident the appearance of the Rain God within the Wing Complex of the Parrot Geoglyph on Mars will make perfect sense.

Besides posing as a parrot the Mesoamerican Rain God can also take on the physical attributes of a jaguar.[11] With this in mind, I returned to the Water Jaguar to reassess its feline features. The Water Jaguar is presented here as a human/feline portrait with an elaborate headdress of lightning spiked feathers. Considering its location right next to the Rain God, it seems highly likely that the Water Jaguar represents the jaguar aspect of the Rain God.

To support this identification, I offer a section of a vase that depicts the Chaac God transforming into his jaguar aspect[12] (Fig. 10.18). He stands wielding his axe and wears a jaguar kilt. His facial features include a single tooth and a spotted jaguar muzzle and ear.

Fig. 10.18. The Rain God as a Jaguar. Detail of Maya vase. Drawing by the author after a photograph by Justin Kerr K1201.

Duplication

As I discussed in chapter 6, this idea of conjoining and duplicating half images was a common practice among many of the cultures of Mesoamerica. To try and explain how this idea developed, here is an image of a personified sacrificial knife carved into a stone block by the Aztec (labeled *a* in Fig. 10.19 on page 190). Notice the blade is almond shaped with facial features carved within. The left-facing head has a cut-off nose, a single round eye, and an open mouth filled with teeth. The next image is of a stone sculpture, also produced by the Aztec, showing a set of similar sacrificial knives; however in this case they are conjoined together to form one single face (labeled *b* in Fig. 10.19). By joining the two facial blades they complete a portrait of Tláloc, the thunderous Rain God of Aztec mythology. Notice the two round eyes, small nose with a cross bar, and a mouth with long hanging fangs.

Here is an example of a jade axe from Guatemala that is in the form of a

Fig. 10.19. Sacrificial knife.
a. Single sacrificial knife. Aztec block.
b. Conjoined sacrificial knives. Aztec stone sculpture.
Drawings by the author.

cut-in-half figure[13] (Fig. 10.20). Notice the figure is sliced in half from the top of the head, right through its face and down to the bottom of its long gown. The partial face has a square eye form, a broad nose, and a large, open mouth with thick lips. The figure has an arm that folds across its chest and has an intertwining infinity emblem in its headdress.

The infinity emblem is similar in shape and design to the sign worn in the headdress of the Mayan goddess Chac Chel, which represents the weaving of the cosmos.[14] She is normally represented as an aged goddess and is associated with childbirth, weaving, rain, and great floods.[15] Both the axe figure and the Rain God formation on Mars can only be truly realized when they are duplicated to restore their missing halves.

Returning to the harbor area there are dark cracks in the surface that run along the faces of the Water Jaguar and the Rain God. I notice that they both run in a northern direction and align with a small crystal-like tower, labeled A in Fig. 10.21. Following the alignment of the same dark cracks in their southeastern path, they both lead to very shiny mounds, labled B and C in Fig. 10.21, at about a 22° angle. These are demarcation markers. They show the viewer where to duplicate the "cut in half" geoglyphic images hidden within the Port Jetty.

Fig. 10.20. Half figure, Guatemala (Jade Axe). Drawing by the author.

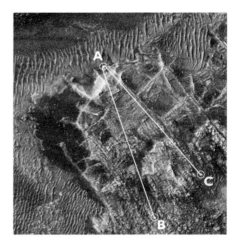

Fig. 10.21. Demarcation markers. Detail MRO HiRISE ESP_067824_1320 (2021). Angles and notations A, B, and C by the author.

Starburst Tower

Heading up to the northern edge of the Port Jetty section of the Wing Complex, I take a closer look at the small tower sticking up from the harbor wall like a mooring post. It is very reflective and has a starburst crest (labeled D in Fig. 10.2).

Here is a detailed image of the Starburst Tower in both the MGS and the HiRISE image (Fig. 10.22). The structure has a spiked, starburst crest that is attached to a thick central rod that sprouts out of a notched, gear-shaped base.

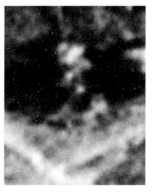

Fig. 10.22. Starburst Tower (labeled D in Fig. 10.2).
Left: Detail MOC S13-01480 (2005).
Right: Detail MRO HiRISE ESP_067824_1320 (2021).

The formation leans to the left, and its spiked, starburst crest may be the result of eons of damage and partial collapse.

Terrestrial Comparison

The overall design of the Starburst Tower resembles a Tesla Tower created by Nikola Tesla in 1891. He built a coil, known as a Tesla Coil, that was housed at the top of a tower. The tower was used as an electrical resonant transformer, which produced high-voltage, low-current, high-frequency alternating-current electricity.[16] His innovative experiments at the Wardenclyffe Plant in Long Island, New York (Fig. 10.23) produced massive bolts of electrical lightning, the kind you see in classic sci-fi movies like *Frankenstein*.

At the Gibbs Farm in Kaukapakapa, New Zealand, there is a sculpture park that showcases the world's largest Tesla Tower called Electrum (Fig. 10.24). Built by Eric Orr and Greg Leyh in 1998, the tower stands 37 feet high and reaches power levels up to 130,000 watts and produces 3 million volts on its spherical top terminal.[17]

The position of the Starburst Tower, which sits on the edge of the Port Jetty, suggests it may have been used for more than a demarcation marker. Its main purpose may have been to be used as a lighthouse or beacon to alert

Public domain, Marc Seifer Archives.

Fig. 10.23 Wardenclyffe Plant with Tesla Tower.
Long Island, New York (1904).

Fig. 10.24. Electrum (Tesla Coil). Kaukapakapa, New Zealand (1998).
Drawing by the author.

approaching ships or aircraft that you have just reached Parrotopia. Or perhaps like a Tesla Tower it was part of a high-voltage apparatus that was used by a Rain God impersonator to produce lightning.

Rulers throughout Mesoamerica would at times impersonate the Rain God by brandishing his lightning axe[18] or carrying buckets of water. Like Tesla some rulers would even depict themselves as producing and controlling lightning. One such example is in the state of Puebla in southeastern Mexico. Situated on the eastern wall of the main chamber of tomb 2004-1, at a site known as Ixcaquixtla, there is a mural showing a ruler controlling lightning[19] (Fig. 10.25). The mural depicts a ruler holding levers and generating sparks and flashes of lightning—like a mad scientist. Considering there are photographs of Tesla sitting in his Long Island laboratory surrounded by lightning bolts, it's not hard to imagine him dressed as the Rain God producing his own brand of lightning.

Fig. 10.25. Ruler posing as Rain God, controlling lightning.
Detail of mural at Ixcaquixtla, Mexico. Drawing by the author.

Chac Mool
(The Guardian of the Water Realm)

Directly below and in front of the large fangs of the Rain God is another right-facing profiled head, labeled E in Fig. 10.2. Fig. 10.26 provides a cropped section of the area surrounding the profiled head. The facial features include a squinting eye and a sharp cheek bone that extends along the nasal bridge. It has a large bar-shaped nasal ornament appearing below the nose, like a dark mustache. The portrait has a large ear form with a possible ear spool, a slightly parted mouth, and chiseled chin. The head wears a flat-topped helmet with a thick, blocky pattern that flairs out the back. The helmet includes a strap that extends down like a chin guard. The helmet rests along the fangs and chest of the Rain God formation (labeled C in Fig. 10.2).

Fig. 10.27 provides a second view of the profiled head in the 2021 MRO HiRISE image of the same area. Although darker in tonality the facial features and headdress are still visible. An analytical drawing of the profiled head is provided in Fig. 10.28.

Terrestrial Comparison

The presence of the flat-toped headdress and the rectangular nose ornament observed within the Profiled Head within the Wing Complex strongly resembles the facial features of the Pre-Columbian god known as Chac Mool, the guardian of the Watery Realm. Due to his strong link to water imagery, Chac Mool was closely associated with Tláloc, the Rain God.[20]

Fig. 10.26. Profiled Head (labeled E in Fig. 10.2). Detail MGS S13-01480 (2005).

Fig. 10.27. Profiled Head (labeled E in Fig. 10.2). Detail MRO HiRISE ESP_067824_1320 (2021).

Fig. 10.28. Profiled Head. Analytical drawing by the author.

There is a fine example of a full-figured Chac Mool sitting atop the Temple of Warriors at the Mayan ruins of Chichén Itzá, which is located on the Yucatan Peninsula in Mexico. Notice the common iconography shared between the Profiled Head on Mars and the Chac Mool portrait (Fig. 10.29). Both head

Fig. 10.29. Chac Mool (profile view). Detail of sculpture at the Palace of the Sculpted Columns, Chichén Itzá. Drawing by the author.

types have a flat-topped headdress that flares out at the back end, a large ear flare, and a rectangular, bar-shaped nose ornament.

Jaguar Paw

Looking toward the east, just beyond the Chac Mool geoglyphic formation, which is at the center of the Wing Complex, is a large square-shaped imprint on the ground. The formation is divided into compartmentalized rooms separated by low foundations that resemble a jaguar's paw (labeled F in Fig. 10.2).

The 2005 MGS image of the paw formation (Fig. 10.30) has a square shape with four dark toepads that radiate along its south and western border. It also has an oval-shaped central pad that is lighter in tone and outlined with a dark line. There is a looped linear pattern that is attached to the northwestern side of the central pad.

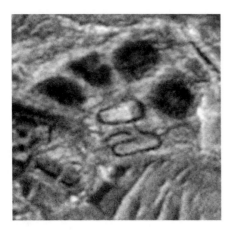

Fig. 10.30. Jaguar Paw (labeled F in Fig. 10.2). Detail MGS S13-01480 (2005).

Fig. 10.31. Jaguar Paw (labeled F in Fig. 10.2). Detail MRO HiRISE ESP_067824_1320 (2021).

In the 2021 MRO HiRISE version of the Jaguar Paw (Fig. 10.31) the imprint of the paw appears to be consistent in size and shape with the earlier image; however, the second and third toe on the western side appear to have suffered an erosional breakdown of the supportive walls that define them.[21]

Terrestrial Comparison

When the Jaguar Paw formation is inverted its compartmentalized divisions become more comparable to the anatomy of an actual feline paw (Fig. 10.32). Notice its cubicle-like divisions form toes and a palm pad.

There are many sculptural examples of feline paws that are comparable to what is seen on Mars. The first example is found within the walls of a site built by the Inca in the fifteenth century on the northern outskirts of the city of Cusco,

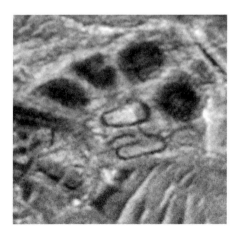

Fig. 10.32. Jaguar Paw (labeled F in Fig. 10.2). Detail MOC 1402185 (2000), inverted.

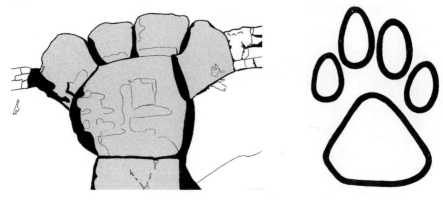

Fig. 10.33. Feline paws.
Left: Puma Paw. Cusco, Peru. Drawing and gray wash by the author.
Right: Lion Paw geoglyph (2010). Chyulu Hills in Kenya, Africa.
Drawing by the author after an image by Andrew Rodgers.

Peru, called Sacsayhuamán.[22] Hidden within one of its composite jigsaw-designed walls is a jaguar paw (Fig. 10.33). Notice the intricately fitted stones take on the shape of a jaguar's paw with four toe pads and a large, central pad.

A second example consists of a set of five linear walls of stacked stone that create a generically shaped lion's paw (Fig. 10.33). The sculpture was created in 2010 in the Chyulu Hills in Kenya, Africa, by one of the world's most recognized land artists, Andrew Rogers.* The formation, which measures 131 feet wide and over 6 feet high, was designed with the aid of the local Masai people.[23] The construction project included over one thousand tribal members who moved two thousand tons of stone to complete the community project. The Masai wanted to make a lasting statement about their traditional lifestyle and leave a permanent "footprint" for generations to come.[24]

Within the rich iconography of the Maya, the presence of a severed jaguar paw is seen as a ceremonial sign signifying kingship. At times the jaguar paw is also worn as a glove or adorned as a crest at the top of a club or royal scepter.[25] The presence of a jaguar paw also denotes a mythological reference to the second Hero Twin known as Xbalanque. He is often characterized by jaguar patches on his chin, arms, and legs.[26]

---

*Andrew Rogers is contemporary land artist who was born in Australia in 1947. He has created the world's largest set of contemporary land art that is titled *Rhythms of Life*. This ambitious project began in 1998 and over the past eighteen years he has created over fifty geoglyphic sculptures that are scattered across sixteen countries and seven continents.

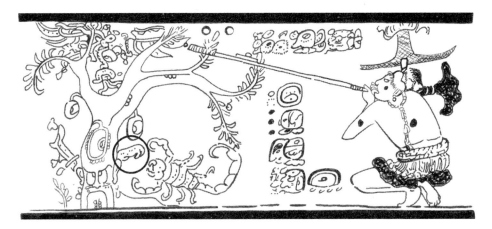

Fig. 10.34. Hero Twin in the guise of a jaguar paw. Notice the jaguar paw peeking from behind the base of the tree (circled). Drawing by the author after a photograph by Justin Kerr K1226.

On a Maya vessel housed in the Museum of Fine Arts in Boston, Massachusetts, is a scene depicting the Hero Twins. The first Hero Twin, Hunahpu is seen in the foreground. He is squatting down and holds a blowgun to shoot a parrot (Seven Macaw) that sits at the top of the World Tree. His brother Xbalanque can be seen peeking from behind the base of the tree, in the form of a jaguar's paws[27] (Fig. 10.34).

Looking again at the compartmentalized cubicles that create the Jaguar Paw, which are separated by low foundations, one can see they resemble the

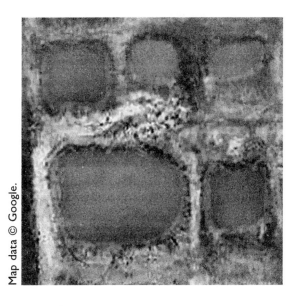

Map data © Google.

Fig. 10.35. Punta Gorda Tropical Fishfarm Aquaculture. Fort Myers, Florida.

segregated aquaculture pools used for shrimp and fish farms. Performing a quick Google Earth search I found the Punta Gorda Tropical Fishfarm Aquaculture in Fort Myers, Florida, that has a very similar paw-shaped grouping of pools (Fig. 10.35 on page 199). Notice the low foundational walls that separate the pools reflect the same paw shape seen on Mars. Considering the idea that the Wing Complex is a port city, the construction of a fish farm makes perfect sense.

Water Processing Plant (The Scarab)

Just beyond the Jaguar Paw and the crumbling bridge-like overpass is a smooth, oval-shaped object that looks totally out of place in this chaotic terrain (labeled G in Fig. 10.2).

In 2002 this area was extensively examined by independent researcher Wil Faust and image analyst James S. Miller in the original Anomaly Hunters Group Study. The team noticed an oval-shaped object with a pipe rising from

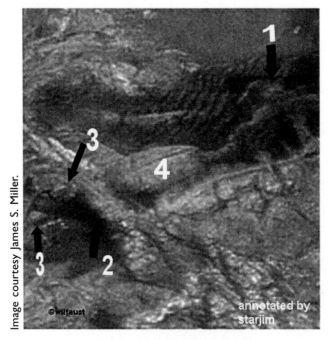

Image courtesy James S. Miller.

Fig. 10.36. Water Processing Plant.
Detail MOC 1402185 (2000).
1. Round object with pipe. 2. The bridge.
3. Ruined beams and buttress.
4. Oval object (scarab).

its center and running off to the right (labeled 1 in Fig. 10.36). They also saw what appeared to be a bridge-shape structure (labeled 2 in Fig. 10.36). Its surrounding area (labeled 3 in Fig. 10.36) has what appears to be the ruined remains of collapsed beams and buttresses that supported the bridge. The area labeled 4 in Fig. 10.36 reveals what Faust thought looked like an Egyptian scarab.[28] He also noticed a distinct line running down the center of its back and the two dark dash marks on its western side, giving it an appearance of a bug's head and wings. The object has linear alignments of rectilinear structures that are similar to long, tubular pipes that are attached to its lower and eastern side, which appear to be half buried in silt and debris. The pipes run parallel to each other and break off into a grid pattern leading to section 1. Faust and Miller propose that this area contains the ruins of a water processing plant for the city. They suggest that section 2 is the intake area and the oval-shaped object in section 4 is the main holding tank, while the round object in section 1 is the processing center.[29]

Here is a close-up of the area surrounding the Water Processing Plant with its attached pipelines as seen in the 2005 MGS image shown in Fig. 10.37. The

Fig. 10.37. Water Processing Plant, labeled F in Fig. 10.2
Detail MGS S13-01480 (2005).

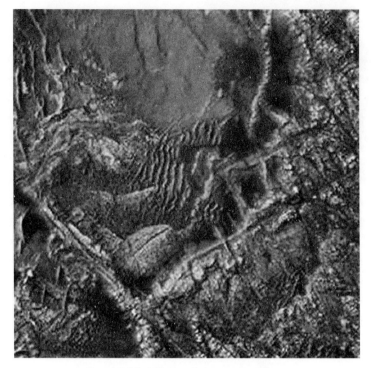

Fig. 10.38. Water Processing Plant.
Detail MRO HiRISE ESP_067824_1320 (2021).

main Storage Tank has two dark dash marks on its western side and a central line that runs across the object from the western to the eastern side. The complex infrastructure is repeated in the MRO HiRISE image shown in Fig. 10.38.

Terrestrial Comparison

There are numerous industrial plants all over the world that process and store water, natural gas, and petroleum products. Searching the internet, I found many comparable storage tanks. Here is a storage tank produced by the Wylie Manufacturing Plant in Texas that resembles the Water Tank observed on Mars (Fig. 10.39).

Wing Complex Comparison

The idea of producing a zoomorphic city in the shape of a parrot's wing seems far-fetched; however as we saw with Evita City in Chapter 6, pictographic

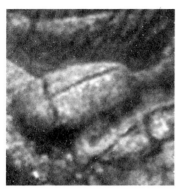
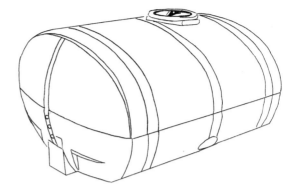

Fig. 10.39. Storage Tanks.
Left: Storage Tank on Mars. Detail MGS S13-01480 (2005).
Right: Storage tank. Wylie Manufacturing Plant, Texas.

designs have come to fruition. The Maya city of Utatlán, which is located near present-day Santa Cruz del Quiché, was designed in the form of an opened-winged Parrot. Its avian form was not realized until a series of aerial reconnaissance studies were performed in the early 1970s by Dwight T. Wallace.[30] After

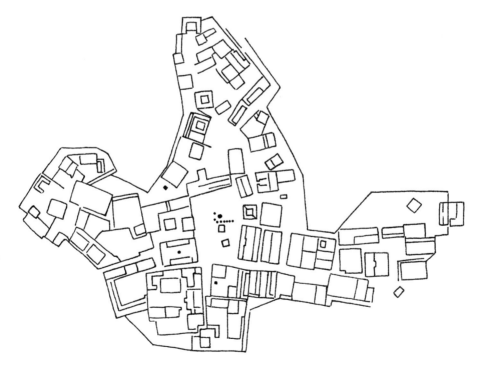

Fig. 10.40. The Utatlán city complex.
Drawing by the author after Robert M. Carmack.

mapping the ruined city was complete the architectural layout of its temples, monuments, and buildings began to take on the contours of a parrot with an upturned wing (Fig. 10.40 on page 203).

The city of Utatlán was an important geoglyphic landmark founded by the early Quiché Maya and is even mentioned in the Popol Vuh.[31] It was the fortress capital of the Quiché Highland state and was originally called Q'umarkaj, "Place of old reeds."[32] The only difference between the parrot-shaped complex in Santa Cruz and the one found on Mars is that the structural formations within Parrotopia reside only in the wing, while the structures that occupy the city of Utatlán reside within the entire bird.

ELEVEN

The Hero Twins and the Turtle of Creation

The Profiled Head

AFTER EXPLORING THE MAIN PARROT GEOGLYPH (labeled A in Fig. 11.1 on page 206) and the pier settlement in the Wing Complex (labeled B in Fig. 11.1), I direct your attention to the adjacent landform just above the Parrot Geoglyph in the MGS image M14-02185 that was released in 2001 (labeled C in Fig. 11.1). Notice the edge of the landform takes on the visible contour of a profiled head looking directly at the parrot.

When the landform is cropped and rotated 90°, the formation resembles a human face with a tonsured head (Fig. 11.2 on page 206). The facial features include a sloping forehead, a profiled eye, and a lid. It has a turned-up nose and a pair of tightly puckered lips above a chin and jaw line. All of these features adhere to the proper size, shape, and orientation of human anatomy.

The profiled head also includes a set of marks on the cheek forming the letters I and L. The forehead and cranial cap appear hairless, while the ear is obscured by a tangle of surface elements that may represent hair. Following a line from the chin to the back of the head, the surface rises abruptly in elevation and becomes darker and highly textured.

In 2013 the Mars Reconnaissance Orbiter acquired a new HiRISE CTX image of the same area with a resolution of 5.0 meters per pixel. The image not only confirmed the anatomical features observed in the Parrot Geoglyph and formations within the Wing Complex, but it also confirmed the facial

Fig. 11.1. Parrotopia Complex.
Detail M14-02185 (2001).
A: Parrot Geoglyph.
B: Wing Complex. C: Profiled Head.

Fig. 11.2. Profiled Head. Detail MGS
M14-02185 (2001).

contours of the Profiled Head (labeled A, B, and C in Fig. 11.3). Along with these geoglyphic and structural formations, this expansive view of the area also included a small egg-shaped mound with an interesting morphology (labeled D in Fig. 11.3).

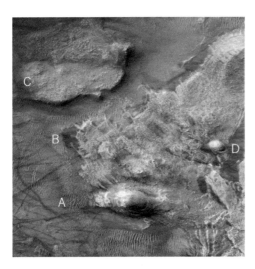

Fig. 11.3. Parrotopia Complex.
Detail MRO HiRISE CTX,
D15_033142_1288_XI_51S054W
(2013). Labeled A–D by the author.
A: Parrot Geoglyph.
B: Wing Complex.
C: Profile Head.
D: Egg-Shaped Mound.

Seven years later, the same Mars Reconnaissance Orbiter MRO HiRISE image that reimaged the Parrot Geoglyph also captured a large swath of territory on its eastern side that includes a great view of the Profiled Head (Fig. 11.4). This MRO HiRISE image was taken in the winter during the early morning with a resolution of 50.4 cm per pixel. The new image confirms all the facial features of the Profiled Head, showing the nose, mouth, and an eye that has an eyeball and eye lids. The "IL" mark can be seen on the cheek, and the tangled hair appears to have decorative braid work. An analytical drawing is provided in Fig. 11.4.

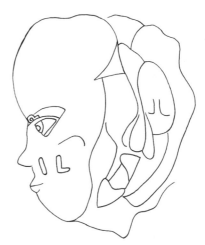

Fig. 11.4. Profiled Head.
Left: Detail MRO HiRISE ESP_067824_1320 (2021).
Right: Analytical drawing by the author.

Cultural Comparison

The Maya often depicted their corn god with a foliated head and corn foliage growing from the side of his head like a corncob. The Maya also represented him with a tonsured head that represented a smooth gourd.[1] In the court of Maya lords, young males that were selected to take care of the temples had the crown of their heads shaved, while the hair around the side of the head and neck remained, mimicking an ear of corn.[2]

The following two glyphs are examples of the tonsured headed corn god that frequently display facial attributes that merge with aspects of the moon goddess[3] (Fig. 11.5). Notice the profiled presentation, the puckered lips, the tonsured head, and the "IL" mark on the cheek. In Maya iconography, these cheek marks signify the corn god's relationship with the moon goddess.[4]

While reexamining the profiled head of the Parrot Geoglyph in Fig. 11.6, I noticed an odd protruding spike attached to the beak that has been tentatively identified as a crest. Beyond its anatomical function, this feathery spike may represent some kind of projectile, possibly a "dart." Notice the arrow shape of the shaft that penetrates the beak and the decorative feather-like feature at the opposite end (Fig. 11.6).

When viewed in a Mesoamerican context, the placement of a dart within the beak of a parrot is most interesting. Its presence is reminiscent of the blow dart that is used by one of the Hero Twins to shoot at the great celestial bird known as Seven Macaw and at times referred to as the Principal Bird Deity.[5]

Fig. 11.5. Corn god with tonsured head and "IL" mark.
Left: Detail of inscription from Tikal, Guatemala.
Right: Detail of sarcophagus of King Pacal, Palenque, Mexico.
Drawings by the author.

Fig. 11.6. Parrot Head. Note the dart in beak.
Detail MOC M14-02185 (2001).

The Popol Vuh records a story where Seven Macaw appears at the time of the creation of the world and steals the Sun. As a result, a pair of brothers known as the Hero Twins vow to kill the bird and free the Sun.[6] A scene from this story is painted on a Maya vessel that depicts one of the Hero Twins (Hunahpu) shooting Seven Macaw with his blowgun (Fig. 11.7). According to the myth, his dart penetrates the bird's beak, knocking him to the ground. With Seven Macaw defeated the Hero Twins were able to secure the Sun and return it to its rightful place.

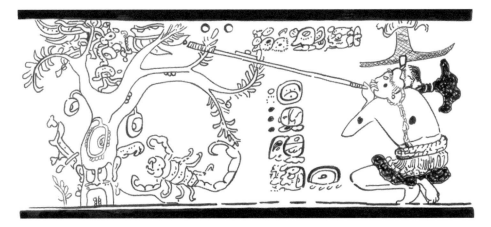

Fig. 11.7. Hero Twin shooting Seven Macaw with blowgun. Maya vase.
Drawing by the author after a photograph by Justin Kerr K1226.

Given the blowing gesture expressed in the puckered lips of the profiled head and the feathered dart sticking through the parrot's beak, I am confident that I have found another match between the mythology of the Maya and the geoglyphs on the surface of Mars. However, these are not the only geoglyphic formations in this close-knit area.

The Turtle of Creation

The egg-shaped mound in the 2013 MRO HiRISE CTX image shown in Fig. 11.3 appears to depict a turtle emerging from an egg (Fig. 11.8). The formation shows a turtle with a triangular head with a small eye and sharp muzzle. You can see its partial body emerging from the egg with an extended leg.

The same 2020 MRO HiRISE image that captured the stunning portrait of the Parrot Geoglyph and the Profiled Head also included the Turtle Mound (Fig. 11.9). The higher-resolution image reveals a circular eggshell that may be a turtle's carapace. It has a triangular head with a small eye and a sharp beak, like that of a snapping turtle. Its body emerges from the egg with its right leg exposed (Fig. 11.9).

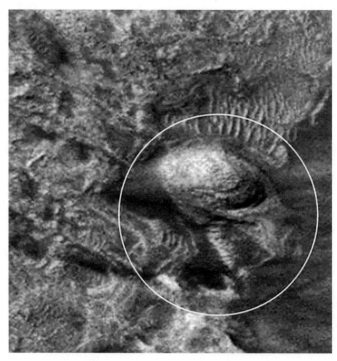

Fig. 11.8. Turtle Mound. Detail MRO HiRISE CTX
D15_033142_1288_XI_51S054W (2013). Location circled by the author.

Fig. 11.9. Turtle Mound.
Detail MRO HiRISE
ESP_067824_1320
(2021).

Terrestrial Comparison

When the Turtle Mound on Mars is compared to a terrestrial turtle hatchling, their physical features are almost identical. Here is a comparison between the Turtle Mound on Mars with a common alligator snapping turtle hatchling (Fig. 11.10). The alligator snapping turtle is one of the larger species of turtles and is found in freshwater habitats throughout the southwestern United States. They can grow to over 35 inches in length and weigh over 220 pounds.[7] It has a large head with a camouflage pattern around its eyes and a powerful jaw. It has skin like an alligator with a rough, ridged shell with three rows of spikes.

Image courtesy U. S. Fish and Wildlife Service Southwest Region (2013).

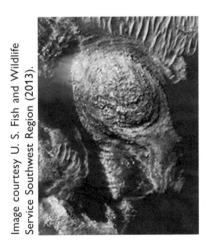

Fig. 11.10. Turtle Mound.
Left: Detail MRO HiRISE ESP_067824_1320 (2021).
Right: Alligator Snapping Turtle Hatchling.

Cultural Reference

In many cultures the turtle is seen as a symbol of water and is sometimes linked to the primordial waters of the cosmos. This model is supported by its anatomy. Looking at the simple construction of a turtle, its outer shell consists of a combination of round and square forms. The heavens are represented in the rounded top half of the turtle's shell (the carapace), while the Earth is represented by the square, flat bottom portion of the shell (the plastron). Its four short, bulky legs act as pillars, enabling the turtle to bear the weight of the cosmos.[8]

The turtle plays an important role in Maya mythology. It symbolizes the Earth floating in the primordial sea of creation[9] and is a representation of rebirth and resurrection.[10] In the Museum of Fine Arts in Boston there is a Maya plate that shows the First Father emerging from a crack (or cleft) in the Cosmic Turtle's shell (Fig. 11.11).

As soon as First Father was reborn, a triangular hearth of three "throne stones" was set up by the gods on August 13, 3113 BCE. These hearthstones were also seen as the three stars of Orion's belt, which were set across the back of the Cosmic Turtle.[11] In the Madrid Codex there is an illustration that

Fig. 11.11. The resurrection of the First Father from the shell of
the Cosmic Turtle. Detail Maya plate, Museum of Fine Arts, Boston.
Drawing by the author.

depicts a sky band with a cosmic turtle carrying the three hearthstones of creation on his back (Fig. 11.12).

Five hundred and forty-two days after the resurrection of the Maya god First Father through the cleft in the Cosmic Turtle, four sky bearers set up the four sides and corners of the cosmos. They then erected the central World Tree,* and together these events signified the beginning of the "Fourth Creation," or "Fourth Sun."[12]

The prominent figure of the "Sky Bearers" was called Pawahtun, and he is often depicted wearing a turtle shell. Fig. 11.13 shows a small sculpture of Pawahtun as a turtle. Notice the "water lily motif" on the shell.[13]

Fig. 11.12. Celestial Turtle. Madrid Codex page 71a. Drawing by the author.

Fig. 11.13. Pawahtun emerging from turtle shell. Maya, ceramic. Drawing by the author after a photograph by Justin Kerr K2980b.

*This is the same World Tree that Seven Macaw later sat in when he was shot down by the Hero Twin Hunahpu.

Fig. 11.14. K'inich Ahkal Mo' Nahb. Detail of stone panel of the Temple XIX pier, Palenque. Drawing by the author after Mark Van Stone.

If we visit the ancient city of Palenque, which is in the Chiapas state of Mexico, there is a stone panel inside Temple XIX that depicts a standing ruler named K'inich Ahkal Mo' Nahb (Fig. 11.14). The partially damaged panel shows K'inich wearing a feathered headdress with an inserted macaw head that includes a small Mars Beast snout attached to its beak. Archaeologists and epigrapher David Stuart suggests that the headdress may have included a turtle, but it is now absent due to the missing pieces of the panel.[14]

His name glyph can be seen on the main tablet in Temple XIX at Palenque (Fig. 11.15). The glyph block incorporates two creatures of the Maya creation mythology, the Cosmic Turtle and Seven Macaw. These two characters appear prominently in the Maya Popol Vuh and are also represented as constellations. Seven Macaw is seen as the Big Dipper, while the Cosmic Turtle represents Orion.[15] Believe it or not, this unlikely pair also has merit in science. According to a paper published in the journal *Evolution & Development* by a group of scientists at Yale University, turtles are closer relatives to birds (archosaurs) than lizards and snakes.[16]

In the composite of glyphs, a turtle shell can be seen on the left side with two Ahu glyphs resting under it. A large macaw head is placed on the right with a water lily draping its cheek. When the composite of glyphs is read together, they read "Great Sun Turtle-Macaw Pool."[20]

Fig. 11.15. Name glyph of K'inich Ahkal Mo' Nahb.
Drawing by the author after David Stuart.

If we return to the Parrotopia Complex on Mars and reflect on the appearance of the parrot and the turtle geoglyphs surrounded by a pool of water, perhaps we are not only seeing the remains of a robust city on another world, but one that could also be called the Great Sun Turtle-Macaw Pool.

As below, so above.

TWELVE

The Anunnaki

George J. Haas and William R. Saunders

A Day with Zecharia Sitchin

BEING BIG FANS OF ZECHARIA SITCHIN, my colleague William R. Saunders and I had read all of his books and anxiously awaited the 2004 launch of his latest, *The Earth Chronicles: Journeys to the Mythical Past*. We hoped to meet him at one of his book signing events and convince him to take a look at some of our discoveries on Mars that supports his research with the Anunnaki. Unfortunately, due to unexpected health issues and unforeseen heart surgery Sitchin's book tour was canceled. Making a quick recovery, the tour was rescheduled as a single event that took place in New York City, titled *A Day with Zecharia Sitchin*. Fortunately for me, I was living in New Jersey at the time and the change in venue allowed me to attend.

The all-day event included an amazing PowerPoint presentation of his work with the Sumerian and Anunnaki and a meet and greet, as well as lunch. I found him to be very pleasant and personable and he took the time to speak with everyone. After lunch we returned to the lecture room for the second half of his presentation. It was an overwhelming event and I remember his last slide of the day was of the Egyptian scale weighing his heart against the feather of the goddess Maat.

A year later, Saunders and I sent Sitchin a copy of our first book, *The Cydonia Codex: Reflections from Mars*, for his review. He kindly responded with a hand typed letter thanking us for our interest in his work and also for mentioning so many of his books. Sitchin passed away on October 9, 2010, at the age of ninety.

The following is an excerpt from our book *The Cydonia Codex*.

The Ancient Texts

Many questions arise out of the discovery of these incredible geoglyphic structures on the planet Mars. How did they come to be? Did an intelligent race of beings evolve independently on the planet Mars and then travel to Earth? Was there a former highly advanced race of beings on Earth who traveled to Mars sometime in the distant past that left us these structures? Or is there another explanation that goes far beyond our solar system and even farther into our past? The answer to all of these questions may lie within the oldest known written records of mankind, records that have been found right here on Earth.

Over the past 150 years the translation of recovered ancient texts—Sumerian, Akkadian, Babylonian, Assyrian, Egyptian, Hittite, and Canaanite—has revealed an incredible story of human history. These texts also provide the story of the times long before man walked the Earth. The volume of these texts is overwhelming. Tens of thousands of clay tablets, found during archaeological digs in the Near East, reveal a whole new chapter of our past.

It is now evident that most of the stories and writings of these cultures had their beginnings in Sumer. Even Biblical tales such as the story of Creation, Adam and Eve, the Garden of Eden, Noah's Ark and the Flood, and even the Tower of Babel all found their roots in earlier Sumerian writings. The Sumerian texts are not the oldest, for they themselves make mention of even older writings, "lost books" from before the Deluge. According to these ancient Sumerian texts, there was a time when only the "gods" were on the Earth; man had not yet been created.

The Sumerian text known as "The Myth of Ea and the Earth" refers to E. A. (who became known as Enki) as the leader of a group of fifty travelers who came to Earth from the planet Nibiru (Planet of the Crossing). This happened at a time when man had not yet appeared on Earth. The spelling of the word used in Akkadian for this group of travelers was *Anunnaki*, which translates as "the fifty who went from heaven to Earth."[1] In Hebrew and in the Old Testament of the Bible the Anunnaki were known as the "Nefilim," which means "those who were cast down."[2]

At the time of these ancient writings the term "Heaven" did not have the meaning it has today; rather it was a reference to the area known as the asteroid belt found between Mars and Jupiter. According to the Mesopotamian Creation text Enuma Elish, our solar system once housed a feminine planet called Tiamat, which was struck by a rogue planet called Nibiru (later renamed Marduk in Babylonian text). This event happened when the solar system was still forming. This collision was so severe it caused the splitting of Tiamat. One portion of Tiamat spun off and became the Earth, while the other portion of the planet broke up into small pieces and became the asteroid belt.

Zecharia Sitchin relates:

The Book of Genesis (1:8) explicitly states that it is this "hammered out bracelet" that the Lord had named "heaven" (shamaim). The Akkadian texts also called this celestial zone "the hammered bracelet" (rakkis) and describe how Marduk stretched out Tiamat's lower part until he brought it end to end, fastened into a permanent great circle. The Sumerian sources leave no doubt that the specific "heaven," as distinct from the general concept of heavens and space, was the asteroid belt. Our Earth and the asteroid belt are the "Heaven and Earth" of both Mesopotamian and biblical references, created when Tiamat was dismembered by the celestial Lord.[3]

The Maya have a similar creation story of how the Earth was formed. To create the Earth, the god Quetzalcoatl and his twin, Tezcatlipoca, attacked a celestial monster known as Tlaltecuhtli that was swimming in the primordial waters of the cosmos. They grabbed the Monster by the left hand and the right foot and broke it in two. With its upper half they formed the Earth, and with its lower half they formed the heavens.[4]

Since the planet Mars is located between the asteroid belt and Earth, the original Sumerian definition of the asteroid belt as heaven is explained by Marius Schneider as the Mountain of Mars, which he says is the intersection of the circle of Earth and the circle of Heaven. Mars is the mandorla! We quote Schneider:

The mountain of Mars (or Janus), which rises up as a mandorla of the Gemini is the locale of Inversion—the mountain of death and resurrection; the mandorla is a sign of Inversion and interlinking, for it is formed by the intersection of the circle of the Earth with the circle of heaven. This mountain has two peaks, and every symbol or sign alluding to this "situation of inversion" is marked by duality or by twin heads.[5]

Just as Schneider reveals this idea of duality encoded within the twin-peaked mountain of Mars, with the twin figures of the Gemini and again in the two faces of Janus, we can find common motifs among the iconography of the Maya. The Maya expressed the idea of duality and inversion by producing "twin heads" within the iconography of their bifurcated masks and composite glyphs.

The Coming to Earth and Mars

The Sumerian story of the Anunnaki coming to Earth and Mars is one that comes to us in large part from the writings of noted author and historian Zecharia Sitchin. The story is laid out in his series of books entitled *The Earth Chronicles* and *The Lost Book of Enki*. This gifted researcher has spent decades translating and interpreting the texts of Earth's most ancient civilizations, as well as chronicling the research and translations of other historians. Although at times his interpretations have received ridicule from some of the scientific community, his claims have continuously been vindicated through new discoveries in science, archaeology, and astronomy.

The story we are about to relate is not one from our imagination but from ancient texts, interpreted by those with the knowledge and ability to do so. The following synopsis begins with a short history of the Anunnaki as recorded by the Sumerians and interpreted in large part by Sitchin; it describes the creation of man as well as the existence and possible purpose of the artificial structures left on Mars.

This astonishing story begins on a world known as Nibiru, the "planet of the crossing," with a conflict between two great leaders who contest the rightful rulership of Nibiru. The Anunnaki ruler was Anu, and his challenger was Alalu, who would attempt to usurp the throne. After losing entitlement to the throne, Alalu left in a spaceship in search of gold, which was a highly prized resource on Nibiru. Gold was important to the Anunnaki because their planet was losing its atmosphere, and after attempts to enhance the atmosphere by stimulating volcanic activity failed, they discovered that only gold could save them. They found that gold could be used to maintain life on their planet by disseminating fine particles of gold into the atmosphere. With the supply of gold running low, Alalu set out to find a supply. It was from outside our known solar system that he traveled to what he called the seventh planet: Ealow (Earth).

The Sumerians counted the planets from the outside in, so as one approaches the Sun from the edge of our solar system, Pluto would be the first planet you encounter. Mars would be the sixth planet and the Earth would be the seventh,[6] Venus the eighth, and so on. Because Venus was in the eighth position in our solar system, an eight-pointed star would denote Venus. In turn Mars would be denoted as a six-pointed star. Besides the warrior god Nergal's association with the planet Mars, Sitchin says that the wild beast-man known as Lahmu was also associated with that planet.[7] Lahmu was often depicted as a naked man with a beard and long hair with six curls. The six curls signify his numeric representation of the planet Mars.

After landing on the seventh planet of our solar system (Earth), Alalu explores the entire planet and soon contacts his home world and informs them that he had found the gold for which he had searched. He declared that his discovery would save a dwindling atmosphere on Nibiru, and for this he should be given the throne.[8]

After Anu agreed to another contest with Alalu to determine rulership of Nibiru, Anu's oldest son, E. A. (Ea/Enki), led a group of fifty Anunnaki to Earth to verify the claims of Alalu. The story and achievements of Ea/Enki are described in one of the longest and best preserved Sumerian narrative poems so far uncovered, named by scholars the "Myth of Enki and the Land's Order." A portion told in the first person describes his arrival on Earth. (Some sections unfortunately are illegible.)

> When I approached Earth, there was much flooding. When I approached its green meadows, heaps and mounds were piled up at my command. I built my house in a pure place. . . . My house—its shade stretches over the Snake Marsh. . . . The carp fish wave their tails in it among the small gizi reeds.[9]

Sitchin has determined that this event took place about 450,000 years ago, during a period when Earth was undergoing an ice age.[10] This date was presented in his book *The 12th Planet*, which was published in 1976 just as NASA was photographing the first image of the Face on Mars. If you recall, it was Richard Hoagland who determined in 1984 that approximately five hundred thousand years ago (on the Mars Solstice) was the last time that the heliacal rising of Earth could have been viewed over the Face on Mars. Is it possible that the Face was built to commemorate this major event, the Anunnaki's arrival on Earth?

Searching the planet, Ea determined that sufficient volumes of gold could be mined from the area we know as South Africa to meet the needs on Nibiru. To handle the task of setting up mining and processing facilities, more Anunnaki were needed, which led to more divisions of power. To decide who should have control, Anu was summoned to Earth.

Anu was the father of two sons, Ea and Enlil. Although it was Ea, the firstborn, who established the original colony on Earth, Enlil was given rulership over the Earth. Ea was to rule over the seas and oversee the mining operations—a decision that did not sit well with him. The laws of Nibiru were such that an heir had to be not only the oldest son but one conceived by a paternal half sister.[11] Sitchin reminds us that in Genesis 20:12 of the King James Bible it records that the father of the Jewish nation, Abraham, also married a paternal half sister.

Although Ea was the oldest son, he had not been conceived by a half sister of Anu. Enlil, however, was, and thus he was considered the legitimate heir. This situation led to a number of confrontations between the brothers that eventually led to wars between the descendants of Ea and those of Enlil. Ea was not the only one upset with Anu's decisions. Alalu, who was the first to come to Earth years earlier and the one who initially discovered gold, was given no title at all. Determined to regain his seat of power on Nibiru, he challenged Anu to the contest he was promised. Anu would once again defeat Alalu, but while celebrating his triumph he was attacked again by Alalu, who bit off and swallowed the testicles of Anu. As punishment for this despicable crime, Alalu was sentenced to exile on the planet Lahmu (Mars). It was here that he would remain alone until his death. Upon arriving at Mars, the pilot of Alalu's spacecraft, Anzu, announced that he would stay with Alalu as his companion and protect him until his death. Anzu said that when the time came, he would oversee the burial of Alalu, which would be a burial fit for a former Nibiruian king.

Anu determined that, in order to transport the gold from Earth, way stations needed to be set up on both the planet Mars and the small moon of Earth. For his service to Alalu, Anzu would be given command over the way station on Mars. Ninmah, a medical officer who was the daughter of Anu and half sister to Ea and Enlil, commanded the next flight from Nibiru to Earth. Her job was to transport medicinal supplies to Earth and begin the process of setting up the way station on Mars. When the spacecraft reached Mars, Alalu was missing, and Anzu was found near death. When he was revived, Anzu told Ninmah of Alalu's death and led the landing party to the sacred place of his burial. Sitchin speculates that to commemorate Alalu's deeds as a former Nibiruian ruler and the first king to be buried on an alien planet, his likeness may have been carved on a large rock, such as the famous Face on Mars.[12]

Could Sitchin be right? Are the remains of the edifice that graced Alalu's tomb the bifurcated mask we see on the Cydonia Face today? Is Alalu's tomb actually in the shape of a split-faced mask, one half being feline, facing Earth, and the other half being humanoid, facing Nibiru? If true, the tomb of Alalu and the Face on Mars are one and the same, and the settlement that the Anunnaki called the "way station" is synonymous with Richard Hoagland's "city complex," located on a vast plain we call Cydonia.

The Settling of Earth and Mars

The Sumerian texts describe how, over time, many more Anunnaki arrived on Earth[13] until they numbered six hundred:

"Assigned to Anu, to heed his instructions, three hundred in the heavens he stationed as a guard: the ways of Earth to define from the Heaven: And on Earth, six hundred he made reside. After all their instructions had ordered, to the Anunnaki of Heaven and of Earth he allotted their assignments."[14] Sumerian texts chronicle Anu's declaration that the six hundred inhabitants of Earth would be known as the Anunnaki, while the three hundred inhabitants of the planet Mars would be called the Igigi.

After much time had passed, Anzu and the Igigi on Mars were given passage to Earth for a respite. During this visit a plan was conceived for Anzu and the Igigi to overthrow Enlil's rulership on Earth. Their plan was foiled, however, when Anzu was defeated in battle by Enlil's oldest son and heir, Ninurta. For his evil deed, Anzu was sentenced to death; his body was to be buried on Mars, just as Alalu's had been. Control over the Igigi and the way station on Mars was then given to Marduk, Ea's oldest son and self-proclaimed heir.

The Creation of Man

As time passed, the burden of mining on Earth became too much for the Anunnaki workers, and a rebellion threatened. Texts such as the Babylonian "The Creation of Man by the Mother Goddess" and the Sumerian "When the Gods as Men" spelled out, in detail, the decision to create a primitive worker and the process by which it was to be accomplished.[15] To serve the needs of the Anunnaki, a "lulu" (the mixed one) was created. It is believed that the first man was the result of the genetic manipulation of Anunnaki DNA with that of an existing Earth creature, probably *Homo erectus*.[16] The Creation of Adamu (Earthling) was the work of Ea (Enki) and his half sister, Ninmah, the chief medical officer (also known as Ninharsag or Ninti). To begin the process, Ea announced: "Blood will I amass, bring bones into being."[17]

The Sumerian text "When the Gods as Men" then goes on to state:

"While the Birth Goddess is present, Let the Birth Goddess fashion offspring.

While the mother of the Gods is present, Let the Birth Goddess fashion a Lulu:

Let the worker carry the toil of the gods. Let her create a Lulu Amelu, let him bear the yoke" . . .[18]

It was Ea's wife who was chosen to carry the first hybrid child, created by Ninki. The text continues: "The newborn's fate thou shalt pronounce: Ninki would fix upon it the image of the gods: And what it will be is "Man.""[19]

And Elohim (deities), as written in Genesis 1:26, said: "Let us make Man in our image, after our likeness."

A similar story can be found with the Maya, wherein a set of heroes on Earth asked their mother to create a race of men who would serve them. They asked the god of the Underworld for a bone of past men to mix with their own blood, which would create a man and a woman to multiply and populate the Earth.[20]

The original humanoid that was created by the Anunnaki, Adamu, was a sterile hybrid and unable to procreate.[21] These hybrids worked as miners in South Africa under the control of Ea and as gardeners in Edin (modern-day Kuwait) under the control of Enlil. The addition of the sex genes by Ea (whose symbol is the helical or entwined serpents) allowed man to procreate. The result of these genetic adjustments allowed the Anunnaki to create three different kinds of workers that were adaptable to different environments. Anthropologists have recognized these three basic races as the Caucasian (white), Mongoloid (yellow and red) and Negroid (brown and black). This angered Enlil greatly, and he expelled Man from Edin. Left on his own, Man began to proliferate at an unexpected rate.

Ea loved to sail the waters, and during his travels he came upon two hybrid females. Overcome by their beauty, he mated and impregnated both of them. One bore him a male child, Adapa (the Foundling), and the other a female, Titi (One with Life). The children were raised by Ea and his wife, Ninki. Adapa was taught by Ea himself and became the first civilized man. Adapa and Titi eventually had children themselves, twin brothers Ka-in and Abael;[22] over time, this new breed of civilized man spread throughout the globe.

The Flood

Meanwhile, back on Mars the Igigi were becoming restless and desired the amenities the Anunnaki had on Earth. When it was learned that Marduk was to marry one of the human females on Earth, they too wished the companionship of the Earthlings. A plot was hatched for two hundred of the Igigi to attend Marduk's wedding on Earth and abduct the human females for their wives. They seized the women and, since life on Mars had become too harsh and undesirable, they demanded to be allowed to marry and live on Earth. Marduk and the others agreed.

As a result of this new decree, Enlil became extremely outraged. He had been angered when the first Earthlings were created, furious when they were allowed to reproduce, and was even angrier at the interbreeding between the Anunnaki and Earthlings that was taking place. This story can be found in the Bible as well, in Genesis 6:1–4.

> And it came to pass, when the Earthlings began to increase in number upon the face of the Earth, and daughters were born unto them, that the sons of the deities saw the daughters of the Earthlings that they were compatible, and they took unto themselves wives of whichever they chose. (Genesis 6:1–2)

> The Nefilim were on the Earth in those days—and also afterward—when the sons of the deities went to the daughters of men and had children by them. They were the heroes of old, men of renown. (Genesis 6:4)

Enlil's anger did not diminish. When it was learned that a catastrophe was approaching the Earth that would cause a worldwide flood, he decided that mankind should not be warned, but perish. Fortunately for mankind, Ea would not allow his creation to be destroyed, and he notified one of his sons by an Earthling, Zuisudra, that he must prepare a ship that could withstand the Deluge. In "The Epic of Gilgamesh" the text has Ea speaking clandestinely to Zuisudra:

> "Man of Shuruppak, son of Ubar-Tutu Tear down the house, build a ship! Give up possessions, seek thou life! Forswear belongings, keep soul alive! Aboard ship takes thou the seed of all living things; That ship thou shalt build—her dimensions shall be to measure."[23]

The legend of the Flood is found in most cultures and religions of the world, including those of the Americas. The Olmec, Maya, and Inca all believed that they lived in the Fifth Age of the Sun, and that the Fourth Age ended with a universal flood.

After the Deluge (possibly around 11,000 BCE), when it was discovered that mankind had survived the calamity, Enlil decided that it was destiny and the will of the "Creator of all things" that man should live. This being the case, the Anunnaki made the decision to endow mankind with the knowledge of farming and animal domestication.[24] Eventually kingship was bestowed upon man. The first recoded civilizations to elevate humans to the throne took place in Sumer around 3800 BCE, subsequently in Egypt around 3100 BCE, and in the Indus Valley around 2800 BCE. As further evidence for Sitchin's time-line, we look to the ruins at Caral in Peru, which have recently been dated to 2600 BCE.[25]

Before this time, it was only the Anunnaki and their descendants who ruled Earth's city-states. The Sumerian texts state that eight kings reigned during the 241,200 years before the Flood. The Greek historian Herodotus wrote that in the First Dynasty of Egypt seven great gods ruled for 12,300 years:

Ptah (Ea)—9,000 years. Ra (Marduk)—1,000 years. Shu—700 years. Geb—500 years. Osiris—450 years. Seth—350 years. Horus—300 years.[26]

The second dynasty of the gods, according to the Egyptian historian Manetho, consisted of twelve divine rulers who ruled for 1,570 years. The first of these was Thoth (Ningishzidda). A third dynasty that lasted 3,650 years was ruled by thirty demigods who were half god, half man—that is, children of the Anunnaki by Earthlings.[27]

Ningishzidda Becomes Quetzalcoatl

The Sumerian records suggest that while in Egypt the Sumerian god Ea was recognized as Ptah and his son Ningishzidda was referred to as Thoth. It was Ea who taught the sacred knowledge of the ancients, including the science of genetics. Marduk, the firstborn son and legal heir of Ea, was so outraged that his brother Ningishzidda, the sixth-born son, was given these secrets and he was not that he vowed to retaliate. Ningishzidda as Thoth ruled over the lands of Egypt and, like his father, adopted the image of twin serpents around a rod as a symbol. The caduceus was Ningishzidda's emblem, an adaptation of which is still used today as the symbol for medicine.

It was around the time that kingship was being presented to mankind in

Egypt that Marduk returned from exile on Mars. Marduk became the Egyptian god Ra and deposed his brother Ningishzidda from lordship throughout Egypt. Historians assign an approximate date of 3100 BCE to the beginning of dynastic rule in Egypt. The theory states that Ningishzidda made his way to the Americas with the Black Headed People (the Olmec) and became known to the Mesoamerican cultures as Quetzalcoatl, "The Feathered Serpent." There are a number of clues to justify this theory.

Most archaeologists now agree that the Olmec were the "Mother Culture" of Mesoamerica civilizations,[28] and the Maya derived almost all the knowledge and mythology from the Olmec. The Maya recorded the date 3113 BCE as the beginning of their calendar, which is not likely a coincidence to the date assigned to dynastic rule in Egypt. It was Quetzalcoatl who bestowed the knowledge of writing and the arts and sciences on the ancient Maya, and therefore the Maya must have absorbed the concepts of a bifurcated writing system whose existence extends to the two-faced geoglyphs on Mars. The central symbol in the Maya culture was the "World Tree." The Sumerian name Ningishzidda means "Lord of the Tree of Life."[29] His symbol was the twin serpents, as was that of the Mesoamerican god Quetzalcoatl. The sacred number of the Maya calendar was 52, which was Thoth's "magical number." Also, Ningishzidda had been taught the science of pyramid building by his father. Sitchin states:

> It is recorded that when Ninurta (Enlil's foremost son) desired that a ziggurat-temple (stepped pyramid) be built for him by Gudea, it was Ningishzidda (Thoth) who had drawn the building plans.[30] The stepped pyramids of Mesoamerica are of course a wonder to all that view them, and they are exactly the type of structure for which Ningishzidda (Thoth) was noted.

Hidden in Plain Sight

What this presentation has shown is that the geoglyphic structures on Mars reflect cultural and religious beliefs that include ancient and present-day societies found on Earth. And these geoglyphic structures on Mars are presented in a form that is found within a symbolic language that once flourished in Mesoamerica.

Perhaps the iconography and religion throughout Mesoamerica evolved out of the remnants of an earlier culture that disappeared from the area due to war or natural disaster. What is more likely is that all the cultures throughout

Mesoamerica and South America are connected in some way to the Anunnaki. The common mythologies shared by the cultures of Mesoamerica and the Sumerians were possibly inherited by these New World cultures through the arrival of Ningishzidda and his followers that brought knowledge to these indigenous cultures. It was probably also Ningishzidda, the sixth son of Ea, that set the foundation for a two-faced and composite writing system that encoded the sacred knowledge of the "gods."

Is it possible that the structures on Mars are the result of the unendorsed workings of Marduk and the Igigi, who secretly recorded the sacred history of man and their relationship to the Anunnaki across the surface of Mars in the form of hidden geoglyphs? Outraged by his father's refusal to grant him the sacred knowledge of the ancients, perhaps Marduk stole the knowledge that he felt was due him. Then, in the form of pyramids and geoglyphic structures, he incorporated this sacred knowledge within the structural design of each building, thereby hiding his blasphemy in plain sight.

Is this the answer? Are the structures on Mars just the remnants of gigantic hieroglyphs, recording the story of a once advanced race long forgotten? We may never know until we have a chance to go to Mars and investigate these structures firsthand, or perhaps until new archaeological information connecting Mesoamerica with Mars is uncovered here on Earth. In the meantime, a collaborative research program must be established between NASA and the archaeological communities to further evaluate this lost heritage. The Cydonia Institute, along with many other researchers, is continuing the investigation, and strong evidence has been found that additional half and bifurcated geoglyphic structures can be found all over the surface of Mars.

Growing evidence suggests that the planet Mars was the recipient of an elegant opus that archives the lost heritage of humankind. It appears that some areas of Mars have been literally transformed into a sacred book of pictographic icons. The artistically fashioned surface features observed on Mars simulate holographic pages that unfold within the expansive terrain. A matrix of geoglyphic structures and complexes illustrate a text that is only obscured by its covert design. Like the enigmatic message recorded in the original Popol Vuh, which states "The original book, written long ago existed, but its sight is hidden from the searcher and the thinker,"[31] perhaps the mystery behind these structures on Mars are also hidden from the searcher and the thinker, but for those of us initiated into its matrix . . . our eyes are wide open.

THIRTEEN

The Maya "Star-War"

The Maya Weekend

IN 1999 I BEGAN ATTENDING meetings of the Pre-Columbian Society at the University of Pennsylvania museum. I also participated in glyph workshops with the group's founder, an expert at deciphering Maya hieroglyphs, John F. Harris. His classes were a real eye-opener to an entirely new world of sign language. Wanting to learn more I became a member and over the next ten years, thanks to Harris, I learned to read and write Maya glyphs.

Diving in headfirst, I attended my first Maya Weekend at the museum in March 2000. It was billed as the 18th Annual Maya Weekend titled, *Portraits of the Maya*. The meeting began on Friday evening with a special lecture by the senior research scholar at the Museum, Christopher Jones. After his talk all guests were invited to a reception area in the Egyptian Gallery with hors d'oeuvres and wine. While making my way through the crowd I had a chance to talk with some of my fellow attendees, and it was not too long until I realized that a lot of them were talking about a new Maya researcher by the name of Simon Martin. It seemed everyone was excited about this new speaker and anxiously anticipating his appearance. Simon Martin was a British epigrapher, historian, and writer and was quickly becoming one of the best-known contributors to the study and decipherment of the Maya script. He was being hailed as a rock star and the next Linda Schele.*

*Linda Schele was an American archaeologist and world-renowned expert in the field of Maya epigraphy and iconography. She played an invaluable role throughout the 1980s and 1990s in the decipherment of many of the Maya hieroglyphs.

Martin was the second to speak at the Saturday meeting, and the Harrison Auditorium was packed to the rafters during his lecture on the "Kings of Calakmul's Golden Age." Like many of the starry-eyed assembly, halfway through his talk, I too became a big fan of this new, powerful titan of the Mayan Brotherhood.

I caught Martin's lecture at the next Maya Weekend in 2001, and in 2008 I got to meet him face to face. I participated in his hands-on workshop on Maya glyph reading, titled "Challenging Text from Tikal." The four-hour workshop was intense, and I found him to be engaging and quite approachable. It was here that I first heard of the Maya city of Naranjo and its mystic connection with the Mars Beast. It was also at his workshop that I first became aware of Martin's book, which he cowrote with Nikolai Grube, titled *Chronicle of the Maya Kings and Queens*. In the book Martin and Grube lay out the history and founding of Naranjo and its many battles with its neighbors. Most importantly they examine a cosmic war that occurred with an unknown place that archaeologists just can't seem to find, a place that I believe is directly related to the planet Mars. It was with his book that I started my journey and one by one began connecting the dots between the Maya and Mars.

Naranjo

The little-known and excavated site of Naranjo, which was first discovered in 1905 by a German-Austrian explorer named Teobert Maler,[1] may hold the records of a lost alliance between Earth and a red planet we call Mars. Located between the Holmul and Mopan rivers in Mexico, Naranjo is surrounded by some of the most important Classic Period kingdoms of Mesoamerica including Tikal, Caracol, Calakmul, and El Perú. Throughout its little-understood history Naranjo engaged in multiple battles with its neighbors and suffered significant defeats.[2] The archaeological records say that the construction of large-scale ceremonial structures began at Naranjo around 546 CE, and it became a great capital that was well established by 600 CE.[3] The city peaked during the Late Classic Period around 700 CE and began its descent around 830 CE.[4]

Beyond these established timelines there are two alternate sources at Naranjo that place its conception during two different time periods. The first date is found on Stela 1 that projects its mythical dynastic founding to an astonishing 896,000 years in the past. The second date is recorded on Alter 1, which places its founding to a more modest time of 22,000 years ago.[5]

Over the past century, despite decades of sporadic looting and massive

damage to its monuments, the site of Naranjo has been extensively studied by prominent archaeologists. Soon after Maler's initial discovery a young
archaeologist from Harvard, Sylvanus G. Morley, began studying and dating
the known monuments of Naranjo. In 1909 he mapped the site and ordered
its architectural development into an Early, Middle, and Late Period. He also
noted long hiatus periods where recorded on monuments that were erected.[6]

A little over fifty years later the great Russian American epigrapher Tatiana
Proskouriakoff, who was instrumental in deciphering Maya hieroglyphs and
proving that historical events were recorded on monuments, became intrigued
with the Naranjo. So much so that in 1962 she encouraged fellow archaeologist Richard E. W. Adams to visit the site and preserve its history by making
latex casts of several of the inscriptions. Over the years his work has proven to
be a valuable resource to archaeologists due to the looting and damage that has
occurred to some of the monuments since his visit.[7]

The next to explore Naranjo and contribute to the research was British
archaeologist Ian Graham, who in 1975 produced his *Corpus of Maya Hieroglyphic
Inscriptions*. The publication included photographs and detailed drawings of each
of the monuments.[8] His work along with other archaeologists ultimately led to
Naranjo becoming part of the World Monuments Watch in 2006.[9]

The Maya identified Naranjo with an emblem glyph that is a composite
of different glyphs positioned in a way that creates a frontal view of a face
(Fig. 13.1). It is topped with a pair of glyphs that act as eyes. The eye form on
the left has a variant form of the star glyph LAMAT[10] as a pupil, and the eye on
the right has a T-shaped "IK" sign[11] suggesting a closed eye. Below the eye forms
are a set of glyphs referred to as an sa/SA sign.[12] It creates a nose with an offset
nostril and large arching mouth with a prominent "X" sign. The attached prefix

Fig. 13.1. Naranjo Emblem Glyph.
Drawing by the author.

on the left side is a "holy" CH'UL/K'UL sign, which consists of a K'an cross sign at the top with blood droplets flowing down.[13]

When the main facial elements of the Naranjo Emblem glyph are compared to an Olmec Dragon mask, a common design motif becomes quite apparent (Fig. 13.2a). Notice the shape of the mouth and the inserted "X" cross-band motif, which denotes a sky realm (Fig. 13.2). The cross band is placed in the mouth to denote the center of the sky.[14] Is this "X" motif a celestial reference to an earlier planetary aspect that was embedded within the iconography of the Olmec Dragon? Notice the X sign in the crescent-shaped mouth of the Naranjo Emblem glyph in Fig. 13.2b. A similar X sign is seen on the apron worn by a Maya ruler K'inich Bahlam in Stele 33 at El Perú. The apron consists of a portrait of the Partition God, who is wearing a cross band embedded within his mouth (Fig. 13.2c).

a

b

c
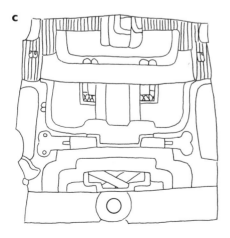

Fig. 13.2. Cross Band Mouth Motif.
a) Olmec Dragon. Note the cross-band sign on the mouth.
b) Naranjo Emblem glyph (detail). Note the cross-band sign on the mouth.
c) Partition God (detail of apron Stele 33—El Perú). Note cross-band "tongue" in the mouth of the Partition God.
Drawings by the author; 13.2a based on image by F. Kent Reilly, III.

The Square-Nosed Beastie

According to hieroglyphic text the kingdom of Naranjo was founded by their patron god, the Square-Nosed Beastie.[15] This was a supernatural creature with obscure origins that go back to the primordial times of creation. Prominent archaeologists such as J. Eric S. Thompson and Michael Coe have also identified the Square-Nosed Beastie as the Zip Monster and Mars Beast, which are all variant signifiers for the planet Mars.[16]

The Mars Beast is presented as a serpentine animal that has a head with a long, curved, fretted snout. Fig. 13.3 offers a few examples. The first is a pair of variant head glyphs of this mystical creature as illustrated in the famous Dresden Codex, which are set within text concerning the revolutions of the planet Mars (Fig. 13.3a, b). Another version can be seen within the sky band that frames the illustrious Lid of Pacal (Fig. 13.3c) and a fourth example comes to us from Alter 1 at Naranjo (Fig. 13.3d).

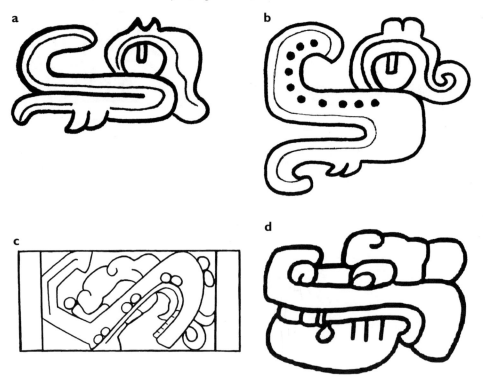

Fig. 13.3. The Mars Beast.
a) Glyph. Detail, Dresden Codex. b) Head. Detail of a descending figure, Dresden Codex. c) Detail of Lid of Pacal sky band. d) Naranjo, Alter 1. Drawing by the author.

Fig. 13.4. The Retrograde Path of Mars. Note the forward to
backward curved pathway of the planet Mars that mimics the
shape of the Mars Beast's snout. Graphic by the author.

There has been much interest and debate among archaeologists as to the odd forward to backward curving snout of the Mars Beast and what iconic significance it may represent.

A Mayan researcher at the University of Texas, Jorge Orejel, believes that the very design of this odd curvature or fretted nose suggests the elongated retrograde motion of the red planet, Mars.[17] Orejel contends that the forward to backward design of the Mars Beast's snout follows the path of the planet Mars that begins by moving in an eastward direction across the sky and then appears to stop (Fig. 13.4). The planet then moves westward, crosses to the other side of the sky, and then appears to stop again. After the second standstill the planet resumes its west to east path forming a curving, backward path.[18]

Stars, Planets, and the Number Six

The ancient cultures of Mesoamerica were aware of at least five of the planets. They were able to recognize Mercury, Venus, Mars, Jupiter, and Saturn. They could see these planets because they are all visible to the naked eye and can be seen moving among the stars.[19]

The Maya created an assortment of logographic and pictographic glyphs that represented most of the known celestial bodies, including the Sun, Earth, the moon, Venus, and Mars. When showing a celestial body, they would sometimes assign it a star sign with a regressive number of points to represent its position in the solar system.

Like the Sumerians the Maya envisioned the order of the planets from the outside in. They saw Mars as the sixth planet, Earth as the seventh planet,[20] and Venus the eighth. Fig. 13.5 on page 234 offers examples of star signs representing Venus, Earth, and Mars.

The planet Mars was not always depicted as a six-pointed star. The Maya

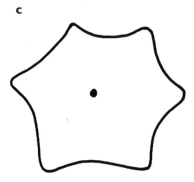

Fig. 13.5. Planetary star signs.
a) Eight-pointed star (Venus). Detail, Codex Zouche-Nuttall, page 21.
b) Seven-pointed star with Spider Monkey (Earth).
c) Six-pointed star. Jade.
Drawings by the author; 13.5b based on photograph by Justin Kerr K7449; 13.5c based on photograph by Justin Kerr K3539.

had a simple numbering system that used dots and bars to count and assign numeral values. A dot equaled one and a bar equaled five. The planet Mars could also be denoted by a set of six dots or spheres attached to a figure within an illustrated scene.

A great example is found at Copán, which is one of the most studied of the Maya cites. It is in western Honduras, right next to the Copán River. The site includes an open-air exhibition park that features a collection of sculptures, one being a feathered Mars Beast, known as Altar O (Fig. 13.6). The serpentine creature has a raised-up fretted snout and six spherical jewels that run along its body. The spheres emphasize the Mars Beast's association with Mars, the sixth planet.

On each end of the sculpture are carved images depicting scenes of the Maya creation story that links the Mars Beast to the time of creation. The triangular panel on the left-end side has a fish and an anthropomorphic frog swimming in the primordial water (Fig. 13.7). In Maya hieroglyphic text, the

Fig. 13.6. Mars Beast with six dots. Altar O, Copán.
Drawing by the author after Alfred Maudslay.

frog is a symbol of birth,[21] and the fish is seen as a symbol of rebirth.[22] The triangular panel on the right side has a full-figured portrait of the Creation Couple (Fig. 13.7), Tepeu and Gucumatz.[23] They stand with a fish next to them that has crosshatching marks on its side indicating that they are in the darkness of the primordial waters.[24]

Fig. 13.7. Side panels of Altar O.
Left: Fish and frog (left panel).
Right: Creation Couple with fish (right panel).
Drawings by the author after Alfred Maudslay.

Fig. 13.8. Mars Beast with six dots.
Detail of Justin Kerr photograph K2774. Drawing by the author.

In the collection of the Indiana University Art Museum in Bloomington, Indiana, there is another example of the Mars Beast's association with the time of creation. A small carved vase shows a Mars Beast wrapped around a star sign and its serpentine body is decorated with six spherical jewel signs (Fig. 13.8.) Within the open jaws of the Mars Beast, you can see the head of Pawahtun, the "old god" emerging from his mouth. Pawahtun was one of the original gods that held up the sky at the time of Creation,[25] and his appearance here again signifies the Mars Beast's connection to Creation Mythology.

Traveling over to Chichén Itzá, which is in the Northern Maya Lowlands, there is a structure known as the Temple of the Jaguar that is thought to have been built between 1000 and 1150 CE. It takes its name from a sequence of jaguars located in front of the structure.

On a large panel found within the Lower Temple of the Jaguar is an image of the Mars Beast that again places its appearance within a prominent role of the creation story. The main figure depicted on the mural is the "old god" known as Pawahtun (Fig. 13.9). He wears a turtle breast plate and an almond-shaped medallion (Fig. 13.9a). Being an aquatic creature, the turtle is a symbol of the primeval waters[26], and the almond-shaped medallion is actually a representation of a zero glyph[27] (Fig. 13.9b). The "old god" stands with two fish nibbling at a water lily sprouting around his headdress, and he is also flanked by a pair of Mars Beasts. The pair of Mars Beast heads display a variant form of the zero glyph[28] used for their eyes, which again denotes the zero time of creation. Looking closer at the Mars Beast on the lower right side, he sits on an offering bowl with six spheres or dots denoting his connection to Mars.

Fig. 13.9. Pawahtun at the time of Creation. Detail of the outer facade of the Lower Temple of the Jaguar (Chichén Itzá). Drawing by the author after Linda Schele.
a) Maya Zero glyph.
b) Maya Zero glyph.
Drawings by the author.

Fig. 13.10. Pawahtun with Mars Beast headdress. Codex Zouche-Nuttall, page 44. Drawing by the author.

Another example of this close and enduring relationship between Pawahtun and the Mars Beast can be seen in the Aztec Codex Zouche-Nuttall* (Fig. 13.10). Here Pawahtun appears to be sleeping while wearing a turtle shell and an elaborate Mars Beast mask as a headdress.

Returning to Naranjo there is another reference to Mars recorded on a stela, which commemorates a new Queen, Lady Six Sky (Fig. 13.11). The monument, labeled Stela 24, marks her arrival in 684 CE from the neighboring city of Dos Pilas. The stela depicts the Queen wearing a jade-netted skirt and holding an offering bowl of sacrificial implements. She stands on a half-naked captive, asserting her great power and dominance.[29]

The surviving text describes her initial actions as ordering "the destruction of buildings" and the reopening of a lost portal to the Otherworld. She had the ability to open portals and reestablish a sacred connection to their ancestors, a connection that was broken by the enemies of Naranjo in the past.[30] Was this Otherworld portal a temporal teleportation "jump room" allowing the shaman of Naranjo to reconnect with their ancestors of Mars?

---

*The Codex Zouche-Nuttall is an accordion-folded book that records the genealogies, alliances, and conquests of several eleventh- and twelfth-century rulers of a small Mixtec city-state located in the highland of Oaxaca, Mexico. It now resides in the British Museum.

Fig. 13.11. Lady Six Sky, Stela 24. Drawing by the author after Linda Schele.

In the glyph block on the left side of Stela 24, Lady Six Sky's name is shown as a profiled head portrait. Notice the bar and dot attached to the back of her head assigning her to the number six, again denoting her association with Mars.

Six Earth Place

Further evidence that the planet Mars is equated with the number six is documented within the inscriptions that record an ancient "star-war" between Tikal and a mysterious, little-known city, identified in local inscriptions as Wak Kab'nal.[31] The title Wak Kab'nal means Six Earth Place and was described as the original homeland of the Square-Nosed Beastie.[32]

According to archaeologists these star-war events were instigated by the appearance of Venus. The Maya had an astronomical ritual of timing their battles to the cycles of Venus, such as its appearance as the Morning and Evening star.[33] Here are examples of the Maya glyphs for Star War and Six Earth Place (Fig. 13.12). The Star War glyph is often presented as an EK' star sign[34] over a yi' shell sign[35] (Fig. 13.12).

The Six Earth Place glyph is written as a dot with a vertical bar, which represents the number six and a Kab' earth sign[36] attached to a nal' place sign.[37]

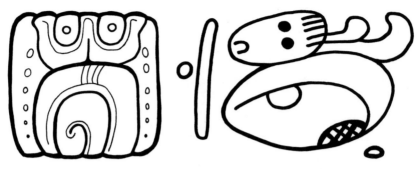

Fig. 13.12. Naranjo star-war and Mars glyph.
Left: Star War.
Right: Six Earth Place glyph (Mars).
Drawings by the author.

This historic city is mentioned numerous times in text found at Naranjo and Tikal. However, it has eluded archaeologists for over a century and as of this publication, the whereabouts of Six Earth Place has not yet been located. Considering the immense time frames related to the founding of Naranjo with its patron god the Square-Nosed Beastie, a mysterious creature associated with Mars, one would presume that Six Earth Place is probably not in Mexico, or even on Earth. I suggest that that this long-lost settlement refers to the sixth planet in our solar system that the Maya considered to be an Earthlike place, ergo Six Earth Place.

The Invasion

The invasion of Earth by some unknown occupants of Mars is presented in plain sight in the Dresden Codex,* which is believed to be one of the oldest surviving books written by the indigenous people of Mexico between the eleventh and twelfth century CE. There are two examples of this celestial invasion in the Dresden Codex, presented as figurative Mars Beast descending from the sky. The first appearance is seen on page 45, which shows an invading army of Mars Beast descending from a sky band (Fig. 13.13).

---

*The Dresden Codex is one of the oldest screen-fold Maya books written in the Americas and dates to the eleventh century. It was rediscovered in the city of Dresden, Germany, hence its present name, and it is currently located in the museum of the Saxon State Library. The codex contains information relating to astronomical and astrological tables, religious references, seasons of the earth, and illness and medicine. It also includes information about celestial bodies, especially Venus and Mars.

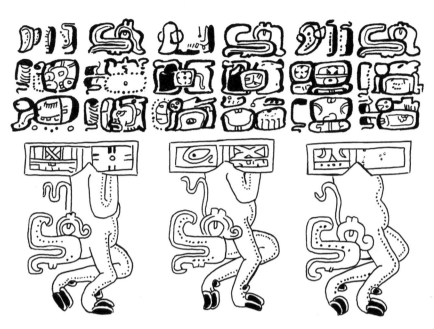

Fig. 13.13. Mars Beast descending from sky band.
Detail, Dresden Codex, page 45. Drawing by the author.

Fig. 13.14. Mars Beast descending from sky band.
Detail, Dresden Codex, page 68. Drawing by the author.

Fig. 13.15. Mars Beast at war. Detail, Dresden Codex, page 45.
Note the axe glyph on left. Drawing by the author.

The second time he appears as a single creature on page 68 (Fig. 13.14 on page 241). On both pages the Mars Beast has the body of a deer, with a short tail and cloven hooves. If you look closely at the set of glyphs running across the top of page 45, there is a set of three Mars Beast glyphs that have axe glyphs next to them. The axe glyph is a reference to the Chaac God, the god of rain and lightning, and the axe also signifies the act of war[38] (Fig. 13.15).

Some may find the presence of the deer's body conflated with a Mars Beast head confusing, but it really isn't. The deer is seen as the bearer of drought,[39] and to the Maya it not only served as a symbol of the hunt and survival, but it was also a metaphor for war.[40] In the Codex Fejérváry-Mayer* a lone deer appears to be confronted and reprimanded by the rain god Tláloc, suggesting his control of the deer and its relationship to rain and the planet Mars (Fig. 13.16).

The Aztec god of rain and lightning known as Tláloc is an equivalent to the Maya God Chaac. This connection of the deer with Mars can be seen again in the Codex Borgia,† where a deer is depicted carrying a six-pointed Mars star across the sky (Fig. 13.17).

---

*The Codex Fejérváry-Mayer is an Aztec book produced in central Mexico. It is one of the rare pre-Hispanic manuscripts to survive the Spanish conquest of Mexico. It is a calendar codex and a divinatory almanac illustrated in seventeen sections. It is made on deerskin parchment designed in a folded accordion-style into twenty-three pages. It is named after Gabriel Fejérváry, a Hungarian collector, and Joseph Mayer, an English antiquarian who bought the codex from Fejérváry. It is currently kept in the World Museum Liverpool in Liverpool, England.

†The Codex Borgia is a Mesoamerican manuscript consisting of thirty-nine pages that were written sometime before the Spanish conquest of Mexico. The codex was brought to Europe during the early Spanish Colonial period and rediscovered in 1805 among the effects of Cardinal Stefano Borgia. The Codex Borgia is presently housed in the Apostolic Library located in the Vatican.

Fig. 13.16. Deer and Tláloc. Detail, Codex Fejérváry-Mayer, page 26.
Drawing by the author.

Fig. 13.17. Deer carrying a six-pointed star. Detail, Codex Borgia, page 33.
Drawing by the author.

Fig. 13.18. Chaac God as the Mars Beast descending from the sky band.
Detail, Madrid Codex, page 2. Drawing by the author.

In the Madrid Codex* there is a partially damaged page that shows a set of full-bodied Mars Beasts descending from the sky. Each figure has a Mars Beast head, and this time the deer's body has been replaced with a human body. Two of them are wielding axes, while a third holds a lightning torch. The axe and the lightning torch are both diagnostic markers for the Chaac God or Tláloc (Fig. 13.18).

At the archaeological site of Copán, located in western Honduras near the border with Guatemala, there is a beautifully carved panel known as the Margarita Panel. A section of the panel shows the Chaac God holding his powerful lightening axe while riding the head of a Mars Beast (Fig. 13.19).

The Chaac God's appearance with the Mars Beast in Maya glyphs is not surprising. Besides utilizing various star points and dots the Maya also used a set of profiled head glyphs of their gods to represent a numbering system from 1 to 19. It just so happens that the number 6 is represented by the profiled head of the Chaac God, and he has an axe sign for an eye.[41] Here is an early

---

*The Madrid Codex (or the Tro-Cortesianus) is one of the four known Mesoamerican codices. The manuscript was created in the late Mayan period somewhere between 900–1521 CE and separated into two parts known as the Troano Codex and the Cortesianus Codex soon after its discovery. The separate codices were not united until 1888. The Madrid Codex is now preserved in the Museo de América, in Madrid, Spain.

Fig. 13.19. Chaac God Riding Mars Beast. Detail, Margarita Panel, Copán.
Drawing by the author.

Olmec version and a later Maya example of the head glyph for the number 6.
Notice both glyphs have an axe sign in their eye, a reference to war and Mars
(Fig. 13.20).

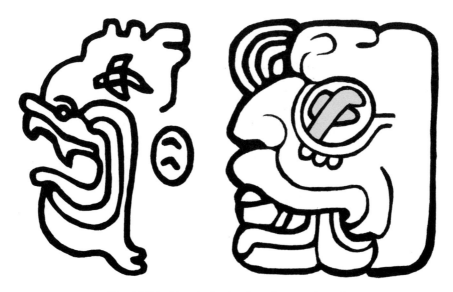

Fig. 13.20. Head glyphs for the number 6.
Left: Olmec. Right: Maya.
Drawing by the author.

It should now be clear that the appearance of the Mars Beast in Maya iconography is not only a symbol of Mars, but it shares a direct connection between the deer and the god of rain and lightning known as the Chaac God by the Maya and Tláloc by the Aztec.

The Battle for Mars

Returning to the Temple of the Jaguar at Chichén Itzá, the walls of the lower chamber are carved with rows and rows of warriors wearing feathered headdresses and carrying strange weapons (Fig. 13.21). The top register of the main panel shows a large circular disk with triangular points projecting around its outer edge. It has a feathered warrior inside that archaeologists call Captain Sun.[42] He is accompanied by two jaguars that form a jaguar throne. Notice on the right side of the main figure is a large Mars Beast head with an elaborate bejeweled snout (Fig. 13.21).

A mural in the upper chamber of the Temple of the Jaguar shows a similar image of a warrior within a star disc, however in this case there are four Mars Beast heads attached to each of its corners (Fig. 13.22). The figure inside the disk is also identified as Captain Sun appearing as an ancestral king wielding a spear thrower and darts, which signifies a Tláloc War.[43]

Fig. 13.21. Rows of warriors with Captain Sun disk at top. SD-5063, drawing. Carved panels, south and west, lower Temple of the Jaguar, ball court. Chichen Itza site, Maya.

Fig. 13.22. Captain Sun with four Mars Beasts.
Detail of northwest mural in Temple of the Jaguar, Chichén Itzá.
Drawing by the author.

Fig. 13.23. Mesopotamia wing and sun disk.
Left: Ashur in winged disk. Detail of wall panel,
Northwest Palace (860 BCE), British Museum.
Right: Figure in Sun disk. Sumerian cylinder seal. Akkad period (detail).
Drawing by the author.

Mesopotamian gods of the Sumerian and Assyrian cultures are commonly depicted in similar sun disks that often appear over scenes of battle.[44] The first example is of the god Ashur depicted here as a "feather-robed archer" holding a bow in a winged sun disk (Fig. 13.23). The second example has a single figure

Fig. 13.24. Aztec warriors.
Left: Warrior with shield with seven dots. Detail, Codex Mendoza, page 67.
Right: Warrior with shield with Mars Beast snout.
Detail, Codex Mendoza, page 67. Drawing by the author.

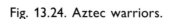

with a bow sitting within a sun disc that has triangular star points that radiate around the disc, much like those seen at Chichén Itzá (Fig. 13.22).

The commonality between the Mesopotamian star disk design and the Mesoamerican version is quite remarkable.

Maya and Aztec warriors are commonly depicted with shields that have different insignia, which represent their military order and alliances. In the Codex Mendoza* there is a pair of warriors brandishing shields that signify both the Earth and Mars. One of the warriors has a shield with seven dots, signifying his representation of Earth, while the other warrior has a shield with only the fretted snout of the Mars Beast, signifying his alliance with Mars (Fig. 13.24).

---

*The Codex Mendoza is an Aztec codex containing pre-conquest history and descriptions of daily life. Named after Don Antonio de Mendoza, who supervised its creation around 1541, it is held at the Bodleian Library at Oxford University.

Fig. 13.25. Nuclear event targeting Cydonia Mensa and Galaxias Chaos.
MOLA map of Mars.

Perhaps these types of shields were used during reenactments of war games and signified which side the warrior would support, either Earth or Mars.

During his research of data collected by NASA of the Martian atmosphere, theoretical plasma physicist Dr. John E. Brandenburg discovered that they had found evidence of a super-abundance of xenon-129 on Mars as early as the first Viking Missions. He also found that gamma ray spectrometry readings taken over the past few years show spiking radiation from xenon-129 on Mars. These are the same type of readings seen on Earth after a nuclear reaction or a nuclear meltdown. Brandenburg compared these types of readings to what was observed at Chernobyl in 1986 and the disaster that happened in Japan in 2011. The unique footprint of xenon-129 indicates an unnatural source for a massive thermonuclear explosion.[45]

Brandenburg believes two nuclear explosions occurred on Mars above the Cydonia and Galaxias Chaos areas of the planet a little over a million years ago (Fig. 13.25). He also believes that the absence of any large craters within each of the exposed sites suggest that the massive thermonuclear event occurred in midair.[46]

If NASA knew about nuclear explosions on Mars, as far back as the Viking era, would they also be aware of the many surviving structures on Mars (such as the famous Face on Mars) that have a striking resemblance to the artwork produced by Mesoamerican cultures?

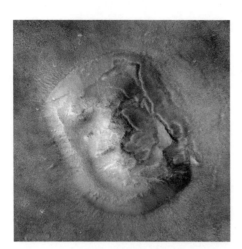 

Fig. 13.26. Bifurcated (two-faced) mask.
Left: The Face on Mars (Humanoid and Feline).
Detail MOC E03-00824 (2001).
Right: Pre-Columbian mask (Feline and human). International Museum
of Ceramics, Faenza, Italy. Drawing by the author.

In the fall of 2005, I had coordinated a meeting with the curator of Pre-Columbian and Native American art at Princeton University, Gillett Griffin. He agreed to look at a manuscript I was working on that explored the bifurcated and composite artwork of Mesoamerican cultures. He invited me to the university and conducted a personal behind-the-scenes tour of the museum. After lunch at the faculty restaurant, we went back to his house to have a look at his private collection of Maya artifacts.

His house was like a mini museum of Pre-Columbian art, and much to my surprise he had quite a few examples of two-faced artifacts. Taking advantage of the opportunity, I asked him what he thought about the humanoid/feline visage of the famous Face on Mars being compared to a bifurcated artwork of the Maya (Fig. 13.26). He immediately went to his library and brought back a book with a photograph of a mountain in Chalcatzingo Mexico that appeared to have a facial image within its cliffside. Griffin thought the odd formation on Mars, like the facial illusion in the book, was nothing more than a "trick of light and shadows." Returning to his library he came back with a video copy of a lecture that was conducted at NASA by his good friend, Linda Schele. My jaw just about hit the floor. I was not aware that Linda Schele had had any contact with NASA. I was totally gobsmacked to learn about this.

According to the timeline, it just so happens that it was only a year before the launch of the Mars Global Surveyor spacecraft (back in 1996) that NASA invited the world-acclaimed Mayan scholar Linda Schele to speak before their scientists. Her lecture was titled "The Universe: Now and Beyond."[47] Considering Brandenburg's revelations of war on Mars and Schele's intimate knowledge of Maya cosmology and iconography, what did NASA wish to learn? And did Schele also get a behind-the-scenes tour of NASA's archives? We will never know. Schele passed away two years later at the young age of 55.[48]

The Spoils of War

On a vase recovered from Naranjo, known as the Rabbit vase, there are two scenes showing the surrender of the god of sorcery and warfare that archaeologists have identified as God L[49] (Fig. 13.27). The highly theatrical scene appears to show the death of Mars.

On the left side of the vase a Rabbit scribe stands on the head of a Mars Beast with a serpent's projecting tongue. The Rabbit scribe holds the belongings of God L, which include a Mars Beast scepter and an owl hat.

Fig. 13.27. The Rabbit Vase. Detail of photograph by Justin Kerr K1398.
Left: Rabbit with Mars Beast staff (detail).
Right: Sun God with Rabbit (detail).
Drawings by the author.

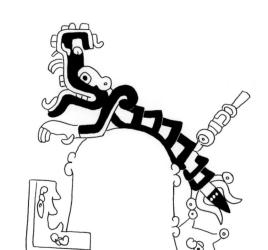

Fig. 13.28. The Death of the Mars Beast. Detail, Codex Zouche-Nuttall, page 51. Drawing by the author.

On the right side of the vase the Sun god sits with the Rabbit scribe on a cushioned throne that rests on top of a Mars Beast. This Mars Beast has a severed deer head impaled on its snout. The Sun God wears a Mars Beast head-dress that has a broken and deflated snout, with a death eye hanging off its tip. Are the characters depicted in this vase acting out the mythological surrender of God L to the conquerors of a dead Mars?

A similar image showing the death of Mars is illustrated in the Aztec Codex Zouche-Nuttall. One of its panels depicts the Mars Beast straddling a mountain (Fig. 13.28) that has been pierced with an arrow, signifying that this place has been conquered or killed.[50] Also notice the mountain is spewing water and the body of the Mars Beast has six scalelike segments.

Heading over to El Perú, which is just to the west of Tikal and located near the banks of the San Pedro River, there was a pair of stelae that once stood side by side within the site's main plaza. The grand pair of stelae displayed portraits of a royal couple that once ruled El Perú (Wak), K'inich Bahlam and his wife, Lady K'abel. The couple are now separated and housed in different locations.

The first stela, which is referred to as Stela 33, is in the collection of the Kimbell Museum of Art in Fort Worth, Texas (Fig. 13.29). It shows K'inich

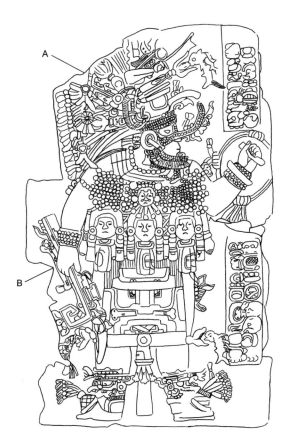

a

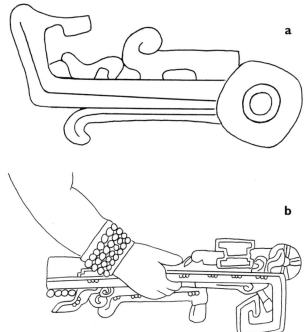

b

Fig. 13.29. K'inich Bahlam (Sun-faced Jaguar). Detail of Stela 33, El Perú.
A. Mars Beast in headdress.
B. Mars Beast with broken snout.
Drawings by the author after Jeffrey H. Miller.

Bahlam wearing a mosaic face mask of a jeweled serpent. He also wears a large headdress of a Waterlily Serpent, which is a supernatural being that symbolizes the earth's abundance of standing bodies of water.[51] Hidden within his feathered headdress is a Mars Beast head (labeled A in Fig. 13.29). His apron contains the face of the Partition God with an X sign in his mouth. This is the same apron that I compared to the design of the Naranjo emblem glyph in Fig. 13.2. In his left hand he holds a round shield, emphasizing his role as a great warrior, and in his right hand he is grasping a Mars Beast scepter with a broken snout (labeled B in Fig. 13.29). The broken snout symbolizes the defeat and death of Mars.

The companion of the pair sits in the Cleveland Art Museum and is labeled Stela 34 (Fig. 13.30). The stela portrays Lady K'abel wearing a six-pointed-star ear spool (labeled A in Figs. 13.30 and 13.31) and a Waterlily Serpent headdress with elaborate blooms of feathers. At the top of her headdress is a Mars Beast head that is oriented in a northern direction with a jeweled snout (labeled B in Figs. 13.30 and 13.31). She stands brandishing a shield in her left hand emphasizing her role as a great warrior and in her right hand she holds a Mars Beast scepter (labeled C in Figs. 13.30 and 13.31). Notice her hand is squeezing the scepter tightly together. In essence she is saying she has crushed the Mars Beast. Her

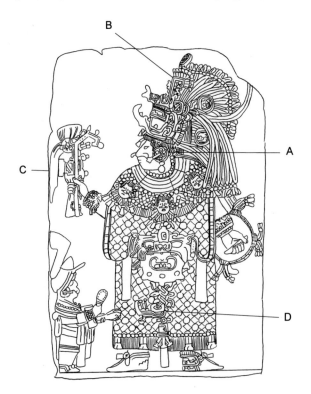

Fig. 13.30. Lady K'abel (Lady Snake Lord). Detail of Stela 34, El Perú. Drawing by the author.

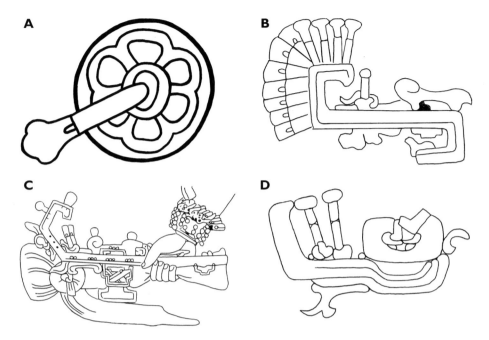

Fig. 13.31. Six-pointed rosette and Mars Beast portraits.
Detail of Stela 34, El Perú.
A. Six-pointed rosette (ear spool). B. Mars Beast in the headdress.
C. Mars Beast staff. D. Mars Beast medallion.
Drawings by the author.

skirt includes a shark-head belt with spondylus shell[52] that has a Mars Beast mask dangling down like a trophy medallion (labeled D in Figs. 13.30 and 13.31).

The broken Mars Beast scepter in Stela 33 recalls the shape of a temple complex found within the heart of the ancient city of Tikal, which is only about thirty-seven miles east of El Perú. Amazingly, the outer contours and the interior design of the main Acropolis at Tikal (Fig. 13.32) project the same shape of the broken Mars Beast scepter held by K'inich Bahlam (Fig. 13.29).

Notice how the placement of the small structures and temples along with its main open plaza creates the profiled image of the Mars Beast with a broken snout. It appears the leaders at Tikal wanted to also celebrate the defeat of the Mars Beast by constructing a set of structures that conformed to the shape of a Mars Beast with a truncated, broken snout.

Further evidence that this Acropolis was a memorial to a defeated Mars Beast can be found within the burial goods of a small temple located at the end of its broken snout. When archaeologists excavated the temple at the tip

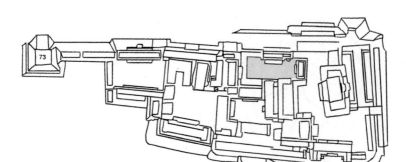

Fig. 13.32. The Acropolis, Tikal. Drawing and gray wash to the
eye feature and notation with the location of Temple 73 by the author.

of the truncated snout of the Acropolis, known as Temple 73, between 1946
and 1965,[53] a beautiful mosaic jade vessel was found (Fig. 13.32 on page 256).
The top of the vessel's lid is crowned with the head of Tikal's twenty-seventh
ruler named K'awiil that Darkens the Sky. He is presented here in the guise of
the Maize God.[54] On one side of the vessel is a single handle in the shape of a
Mars Beast, the same creature from which the Acropolis takes its zoomorphic
form (Fig. 13.32). Another diagnostic clue is that this vessel is identified with
the same Tikal leader, K'awiil that Darkens the Sky, that proclaimed he was the
great warrior that attacked and defeated Six Earth Place,[55] which is Mars.

Fig. 13.33. K'awiil that Darkens
the Sky as Maize God with Mars
Beast handle (Jade Vessel).
Drawing by the author after a
photograph by Justin Kerr K4887.

I find this ruler's claim to be very interesting. Knowing what we know now of Dr. Brandenburg's revelations about the existence of xenon-129 in the Martian atmosphere, was K'awiil that "Darkens the Sky" impersonating the warrior that dropped the bomb on Mars, destroying both the Utopia and Cydonia cities? What better way to express your victory over your nemesis other than building a destroyed Mars Beast monument with a broken snout.

Reenactments

Maya lords and kings saw themselves as part of a divine covenant[56] and would reenact political and military achievements. They would also present themselves as actors who participated in staged mythologies. In some cases, the king would commemorate the same event over and over by rebuilding structures, one on top of the other, or even erecting new ones to portray themselves as various gods.[57] One of the most frequent displays of power was to erect a figurative Stela that presents the king as the World Tree. His body was seen as the trunk and branches of the tree as he held the Double-Headed Serpent Bar in his arms, and the Principal Bird Deity would be set within the summit of his headdress. Archaeologists believe these metaphorical World Trees present the king as "the ambient source of life and the material from which humans constructed it."[58] Some of the elite leadership would also present themselves as the Mars Beast by wearing a partial mask of its fretted snout (Fig. 13.34).

The Maya ball court made a great stage for reenactments. It was used as a metaphor for battle and as a setting for ceremonial decapitations and sacrifice.[59]

Fig. 13.34 Mars Beast masks.
Left: Glyph N. Zapotec.
Right: Maya king wearing a Mars Beast snout.
Detail of Jade plaque. Teotihuacan, Mexico. Drawings by the author.

Fig. 13.35. Reenactment—The Celestial Game of Creation Maya Ball Game: Hero Twins with Deer and Bird Headdress. Drawing by the author after a photograph by Justin Kerr K1209.

In the Chrysler Museum of Art in Norfolk, Virginia, there is a Maya vase showing a pair of ballplayers. They are dressed as the original Hero Twins wearing bird and deer headdresses, reenacting the original Ball Game as recorded in the Popol Vuh (Fig. 13.35).

The Aztec staged an interesting military practice that archaeologists believe originated deep in their distant past, known as the War of Flowers. It was a sacred war game that the ruling class planned with the consent of their rivals. These ritual battles were seen as reenactments of the cosmic struggle between light and darkness. Sometimes they would re-create the mythological conflicts between Quetzalcoatl and his dark brother Tezcatlipoca (Smoking Mirror), or they would stage the quarrels that took place between the Hero Twins and the Lords of the Underworld.[60] These orchestrated battles were solely conducted in an effort not to kill, but to capture prisoners and reserve them for sacrifice.[61]

During these military campaigns the elite warriors would dress as eagles and jaguars. The warriors believed the powers and strengths of these animals would be given to them during battles. The eagle warriors would wear a feathered costume featuring wings and a head cover in the shape of an eagle's head with a beak. Besides assuring the defeat of the enemy, one of the main purposes of the eagle warrior was to capture prisoners and nourish the Sun god with their fresh blood and sacrificial hearts.[62] The jaguar warrior saw themselves as

Fig. 13.36. Aztec eagle and jaguar warriors before the Sun.
Florentine Codex (Book II). Drawing by the author.

a representation of the god Tezcatlipoca (Smoking Mirror). They wore a full-bodied jaguar outfit and carried a war club and attempted to also capture the enemy.[63] In the Florentine Codex* there is an image of eagle and jaguar warriors poised in an attack stance before the sun (Fig. 13.36).

There are massive battle scenes depicted on murals at Chichén Itzá and Bonampak. I will start with one of the murals found at the Bonampak, which is located near the Usumacinta River in the Mexican state of Chiapas. In Room 2 of Structure 1 there is a large mural that depicts what archaeologists have called the greatest battle scene ever produced by the Maya[64] (Fig. 13.37 on page 260). The colorful mural shows warriors engaged in a staged battle, where each side wears similar costumes that represent their military stations. The defeated warriors are presented wearing costumes with an avian element, while the victors are all wearing costumes with a feline motif. At the center of

---

*The Florentine Codex is a sixteenth-century manuscript produced by Franciscan friar Bernardino de Sahagún in partnership with his Aztec students between 1545 and 1590. It consists of 2,400 pages organized into twelve books with over 2,000 illustrations drawn by native artists providing vivid images of this early period.

Fig. 13.37. The Battle Mural. Detail of mural in Structure 1, Room 2, Bonampak. Drawing by the author after Linda Schele.

the scene are two warriors wearing jaguar headdresses. The lead figure stands holding a large pelt-covered spear while subduing a captive by a lock of his hair.

In Chichén Itzá, Mexico, there is a comparable battle scene on the southwest wall of the upper Temple of the Jaguar. The mural shows an active battle scene with warriors carrying spears and brandishing circular shields. Like other murals seen in the Temple of the Jaguar the battle takes place around a large disk that includes a now familiar Captain Sun positioned inside (Fig. 13.38).

Fortunately, these murals at the Temple of the Jaguar were painstakingly copied in the early 1900s by a young archaeologist, Adela Breton, who besides being an archaeologist was also a talented artist. Today the murals are almost entirely gone; they have faded, and portions have even flecked way. Due to the condition of the murals, public access to the rooms is "off limits." Despite the lack of access to these murals, scholars, archaeologists, and the public should be thankful for her fine work, because today we are all able to review and interpret what was originally painted on these walls.[65]

While researching Breton's work at Chichén Itzá I came across a little-known watercolor painting of hers that preserves a battle scene that the public has never seen before. Located in the inner chamber of the upper Temple of the Jaguar, on

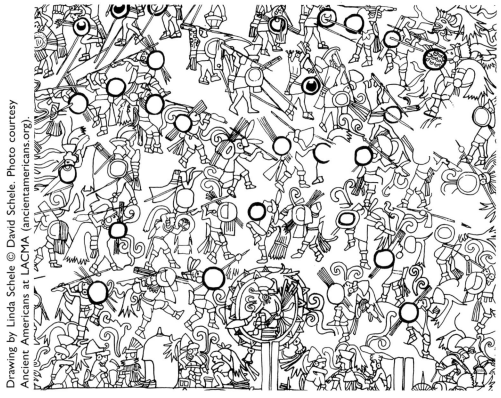

Fig. **13.38.** Captain Sun Disk surrounded by warfare. Detail of southwest panel, upper Temple of the Jaguar. SD-5054, drawing. Chichen Itza site, Maya.

the south wall of the Great Ball Court in Chichén Itzá is a mural depicting a group of warriors fighting on the ground while a sky battle between Sun Captain and a Feathered Serpent Rider commences above them (Fig. 13.39 on page 262). The Serpent Rider appears to be coming in or out of a quadripartite porthole. Does this mural, which is locked away in the upper Temple of the Jaguar, depict a forgotten memory of an ancient "Star-War" battle between Earth and Mars?

As we have seen throughout this study the Maya erected monuments showcasing ceremonial reenactments of their leaders. They were ascending to kingship disguised as gods and recreating military conquests, while others were involved in sky battles that may have occurred almost a million years in the past. These events were repeatedly recorded in their books and on stone monuments or produced in highly orchestrated murals.

This idea of recording historic reenactments on murals and monuments is nothing new. The historic events recorded and reenacted by the Maya and Aztec are no different from all the memorials that we erect to honor our

Fig. 13.39. Sky battle between a Sun Captain and a Feathered Serpent Rider. Detail of wall mural, inner chamber of the upper Temple of the Jaguar, south wall of the Great Ball Court. Chichén Itzá, Mexico. Drawing by the author after Adela Breton.

presidents or glorify our triumphal battles of the past. The National Mall in Washington, DC, is full of them.

Visitors to the National Mall can pay homage to every war that the United States has engaged in from World War I to the controversial Vietnam War. Like the Maya many of these works were produced by descendants of each war's victors and were often erected decades or sometimes even hundreds of years after the event. If you wish to pay homage to a past president you can start at the Washington Monument and take a quick walk over to the Jefferson Memorial, the Lincoln Memorial, and many, many others. Following the history of these massive monuments you can even find evidence of a Mayan-like hiatus period.

It just so happens that during the construction of the Washington Monument in 1848 all work stopped just six years after it had begun because of the lack of funds. As a result, the building of this giant Egyptian-style obelisk suffered a hiatus period of twenty-three years where no work was done. The project was not finished until 1888.

Beyond all this monument building there are numerous history buffs across the country that get together every year, like a platoon of Maya warriors, and dress up in authentic uniforms and reenact medieval jousting tournaments, while others engage in attacks on the Alamo or stage all-out Civil War battles between the North and South. A more passive version of this tradition is achieved when all the citizens of the United States celebrate Independence Day on the Fourth of July. They sit back in the comfort of their lawn chairs and watch exploding fireworks of beautiful colors, which simulate the "bombs bursting in air" as mentioned in the National Anthem.

A country-wide homage to the past is also celebrated every Thanksgiving when everyone participates in a reenactment of the first meal shared between the colonists and the Indigenous people of the New World. As a tribute to faith, on Good Friday, all across the Christian world, many of the faithful dress up as the savior Jesus Christ and carry a large cross through the streets of their hometown, reminding people of the true meaning of Easter.

There have even been reports of hardcore Star Wars fans battling in the streets. One such event happened in Washington, DC, in 2013, where fans dressed as Darth Vader and Storm Troopers fought their way through the streets against numerous Luke Skywalkers and Hans Solos on their way to the Capital building. They were on a quest to promote the acceptance and funding for a new Museum of Science Fiction.[66]

Not far from the clatter of hobnail boots and clashing light sabers on the Capital steps, there is another lesser-known memorial on the other end of Constitution Avenue. Situated in a peaceful elm and holly grove in the south-west corner of the National Academy of Sciences is a memorial that was not erected to immortalize a great battle or presidential hero, but to celebrate a larger-than-life man of science, Albert Einstein. The massive sculpture shows Einstein sitting cross-legged, in casual dress, holding a paper in his lap with an assortment of mathematical equations. Engraved on the bench on which he sits is one of his learned quotes: "The right to search for truth also implies a duty; one must not conceal any part of what one has recognized to be true."

I think Einstein's words make it clear that if we search for truth, we should not hide its supportive facts. We know that it is true that the Maya erected many of their monuments to either document the ascension of their rulers as gods or align themselves with military conquests that occurred in the far and distant past. And we also know that it is a fact that they believed they were performing truthful reenactments and constructing memorials of events that really happened.

This entire story of the Maya having knowledge of an ancient settlement on Earth that had contact with some unknown outpost on Mars, which resulted in a devastating "star war," would seem absurd to almost anyone. That would be true, if it were not for the remains of pyramidal structures on the surface of

Mars. I'm not saying that the Maya lived on Mars and built these structures or ever engaged in any form of Martian warfare that occurred almost a million years ago, but what I am saying is that for some reason, someone implanted this amazing story into their lexicon and creation mythology.

I believe the hieroglyphic text that has been left to us at Naranjo and Tikal and the surrounding settlements record a historic legacy shared between two worlds—a legacy that has been totally overlooked by every archaeologist that has ever studied these records. Perhaps they might want to reexamine the entire opus of Mesoamerican text throughout the region of Naranjo and its neighbors and follow the scattered breadcrumbs they left us. Because I believe that if followed, those breadcrumbs will lead to the doorstep of our point of origin. And like that lost book mentioned in the Maya's Popol Vuh, an original book that was "written long ago, existed but its sight is hidden from the searcher and the thinker,"[67] we might find that unobtainable dwelling of the gods, that "temple not built by hands,"[68] hidden somewhere on the sixth planet of our solar system, a place we call Mars.

Acknowledgments

I would like to extend my sincere thanks to Jon Graham and the publishing staff at Bear & Company along with Inner Traditions and acknowledge Cliff Dunning for reacquainting me with Richard Grossinger, who has been instrumental in the editing and publishing of this book. I am forever indebted to James S. Miller and Dr. John E. Brandenburg for their review of early chapters of this book and contributing their thoughtful forewords. Special thanks are sent to William R. Saunders and Michael Dale for their great insights and geological analysis of these anomalous Martian features. I highly appreciate and thank Ananda L. Sirisena for his critique of early drafts of the book and his knowledge of Carl Sagan. Many thanks go out to Amelia Joy Cole, Phil Hart, Gary Leggiere, Greg Orme, Marc Romanych, Neville Thompson, Evan Sanchez, Javed Raza, Robert Siegel, Amber Ramhorn, Vik Muniz, Dr. John E. Brandenburg, James S. Miller, and Richard and Linda Smith for allowing me to use their photographic images. I also wish to thank the U. S. Fish and Wildlife Service, the LIFE Picture Collection/Shutterstock, and the Los Angeles County Museum of Art for the use of their images. My immense respect and gratitude go out to Justin Kerr for providing the public with access to his enormous photographic archives of Mesoamerican artifacts, a source utilized for many of the illustrations in this publication. I am most grateful to NASA and the ESA for providing the public with such an amazing archive of Martian images.

Finally, I would like to thank my loving and supportive family who have withstood the time diverted from them during the research and writing of this book over the past ten years.

Notes

Foreword.
The Question of Objective Existence

1. Andrew Douglas Reinhard, "Archaeology of Digital Environments, Tools, Methods, and Approaches," (Ph.D. diss., University of York, Archaeology, November 2019), 14.

Preface

1. Jane Boutwell, "Anonymous Art," *New Yorker*, November 14, 1964, 49.
2. Randolfo Rafael Pozos, *The Face on Mars: Evidence for a Lost Civilization?* (Chicago Review Press, 1986).
3. George J. Haas and William R. Saunders, *The Cydonia Codex: Reflections from Mars*, (Berkeley, CA: Frog Ltd, 2005).
4. George J. Haas and William R. Saunders, *The Martian Codex: More Reflections from Mars*, (Berkeley, CA: North Atlantic Books, 2009).
5. Mars Viewer, MOC S13-01480, "Repeat Layered Material and Rectilinear Ridges in M14-02185," Arizona State University Mars Space Flight Facility website (Mars Orbiter Camera), December 15, 2005.
6. M. A. Dale et al. "Avian Formation on a South-Facing Slope Along the Northwest Rim of the Argyre Basin," *Journal of Scientific Exploration* 25, no. 3 (September 10, 2011): 515–538.
7. Erica Phillips, "Earthlings Look for Signs in New Photos of Mars," *Wall Street Journal* CCLX, no. 43, August 21, 2012, 1.

Introduction. Remote Sensing and Learning to See
Past the False Image

1. Haydon F. Stansbury, *Military Ballooning During the Early Civil War* (Baltimore: Johns Hopkins University Press, 1941), 5–15.

2. Mark Dorrian and Frédéric Pousin, eds., *Seeing from Above: The Aerial View in Visual Culture* (London: I. B. Tauris & Co., Ld., 2013), 47.

3. Adam Begley, *The Great Nadar: The Man Behind the Camera* (New York: Tim Duggan Books, 2017), 126–127.

4. Oriental Institute, "Persepolis and Ancient Iran."

5. Anthony F. Aveni, "Solving the Mystery of the Nasca Lines," *Archaeology* 53, no. 3 (May/June 2000): 26.

6. Reader's Digest Association, *The World's Last Mysteries* (Pleasantville, NY: Reader's Digest Association, 1978), 182.

7. United Press, "U.S. Flier Reports Huge Chinese Pyramid in Isolated Mountains Southwest of Sian," *New York Times*, March 3, 1947, 3.

8. Yanek Mieczkowski, *Eisenhower's Sputnik Moment: The Race for Space and World Prestige* (Ithaca, NY: Cornell University Press, 2013), 11–12.

9. Joseph A. Angelo, *Encyclopedia of Space and Astronomy* (New York: Infobase Publishing, 2014), 489.

10. Charles P. Vick, "LACOSSE/ONYX: Radar Imaging Reconnaissance Satellite," Global Security website, July 2005.

11. Kevin Manney, "Tiny Tech Company Awes Viewers," *USA Today*, March 21, 2003, 1B, 8.

12. Yasha Levine, *Surveillance Valley, The Secret Military History of the Internet* (New York: PublicAffairs, 2018), 99, 100.

13. Wall Street Journal, "Google Acquires Keyhole: Digital-Mapping Software Used by CNN in Iraq War," *Wall Street Journal*, October 27, 2004.

14. Alex Turnbull, "The Great White Pyramid of China," Google Sight Seeing website, August 20, 2007.

15. Tasha Shayne, "The Enigma of the Lost Chinese Pyramids of Xi'an," Gaia website, November 12, 2019.

16. Owen Jarus, "Nazca Lines of Kazakhstan: More Than 50 Geoglyphs Discovered," Soul:Ask website, September 26, 2014.

17. Jarus, "Nazca Lines of Kazakhstan."

18. John Noble Wilford, "On the Trail from the Sky: Roads Point to a Lost City," *New York Times*, February 5, 1992, A1.

19. Owen Jarus, "Mysterious Symbols in Kazakhstan: How Old Are They, Really?" Live Science website.

20. Wilford, "On the Trail from the Sky," A1.

21. Owen Jarus, "Visible Only from Above, Mystifying 'Nazca Lines' Discovered in Mideast," NBC Science News website, September 15, 2011.

22. Robin George Andrews, "The Middle East Is Dotted with Thousands of Puzzling Kite-Shaped Structures," Atlas Obscura website, January 22, 2019.

23. Heather Pringle, "Satellite Imagery Uncovers Up to 17 Lost Egyptian Pyramids," *Science* website, May 27, 2011.

24. Sarah Parcak, *Satellite Remote Sensing for Archaeology* (New York: Routledge, 2009), 110.

25. Nicholas St. Fleur, "Desktop Archaeology: In the Saudi Desert, 400 Gates to the Past," *New York Times*, Oct 24, 2017, D2.

26. Brian Dakss, "Old Man of the Mountain Collapses," CBS News website, May 3, 2003.

27. John P. Levasseur et al., "Analysis of the MGS and MRO Images of the Syria Planum Profile Face on Planet Mars," *Journal of Space Exploration* 3, no. 3 (December 30, 2014): 12.

28. Erin Brodwin, "Researchers Won an Award for Figuring Out What Happens in the Brains of People Seeing Jesus in Toast," *Business Insider* website, September 24, 2014.

29. Jill C. Tarter, *Alien Encounters*, Season 1, episode 1, "Part 1: The Message," aired March 13, 2012, on Discovery Channel.

30. Yasuyuki Mamiya et al., "The Pareidolia Test: A Simple Neuropsychological Test Measuring Visual Hallucination-Like Illusions," *Plos One* 11, no. 5 (2016), 5.

31. Longman, et al. "Association of Medical Officers of Asylums and Hospitals for the Insane (London), Medico-Psychological Association of Great Britain and Ireland, Royal Medico-Psychological Association, Harvard University," *The Journal of Mental Science* 13 (1868), 235–238.

32. Steven Goldstein, "Watch What You're Thinking! The Skeptic's Toolbox II Conference," *Skeptical Inquirer* 18, no. 4 (Summer 1994): 345–350.

33. Jennifer Ouellette and Ars Technica, "Why Humans See Faces in Everyday Objects: The Ability to Spot Jesus' Mug in a Piece of Burnt Toast Might Be a Product of Evolution." *Wired* website, July 14, 2021.

34. Bob Dylan, "Subterranean Homesick Blues," recorded January 14, 1965, released as a single by Columbia Records, catalog number 43242, March 8, 1965.

35. Donald M. Anderson, *Elements of Design* (New York: Holt, Rinehart & Winston, 1961), 36.

36. Ronald Greeley, *Introduction to Planetary Geomorphology* (New York: Cambridge University Press, 2013), 26.

37. Anderson, *Elements of Design*, 36.

38. Anderson, *Elements of Design*, 38–39.

39. Tom Van Flandern, "Preliminary Analysis of April 5 Cydonia Image from the Mars Global Surveyor Spacecraft," Meta Research website, April 10, 1998.

40. John P. Levasseur, "Hypothesis," Artifacts on Mars website, 2000.

41. Hugh G. Gauch and Hugh G. Gauch, Jr., *Scientific Method in Practice* (New York: Cambridge University Press, 2003), 11.

42. Christopher Jobson, "WISH: A Monumental 11-Acre Portrait in Belfast by Jorge Rodríguez-Gerada," Colossal website, October 20, 2013.

43. Daisy Wyatt, "Eleven Acre Land Art Unveiled in Belfast," *Independent* website, October 18, 2013.

44. Michael Greshko, "See Newly Discovered Ancient Drawings in Peru Desert," *National Geographic* website, April 5, 2018.

45. John F. Ross, "First City in the New World? Peru's Caral Suggests Civilization Emerged in the Americas 1,000 Years Earlier Than Experts Believed," *Smithsonian Magazine* 33, no. 6 (August 2002): 64.

46. Ruth S. Solis, et al., "Dating Caral a Pre-Ceramic Site in Supe Valley on the Central Coast of Peru," *Science* 292, no. 5517 (May 2001): 723–726.

47. Stanislav A. Grigoriev and Nikolai M. Menshenin, "Discovery of Geoglyphs on the Zjuratkul Ridge in Southern Urals," *Antiquity* 86, no. 331 (March 2012).

48. Personal conversation with J. P. Levasseur, December 2015.

49. Carl Sagan, *Cosmos*, episode 12, "Encyclopaedia Galactica," aired December 14, 1980.

50. K. Morishima, et al., "Discovery of a Big Void in Khufu's Pyramid by Observation of Cosmic-Ray Muons," *Nature* 552 (2017) 386–390.

51. Chris Perez, "Scientists Discover Hidden Chamber in Great Pyramid," *New York Post* website, November 2, 2017.

52. Jean-Pierre Protzen and Stella Nair, "Who Taught the Inca Stonemasons Their Skills? A Comparison of Tiahuanaco and Inca Cut-Stone Masonry," *Journal of the Society of Architectural Historians* 56, no. 2 (June 1997): 146–167.

53. Ella Morton, "The Strange Story of Australia's Mysterious Marree Man," Atlas Obscura on Slate website, January 26, 2015.

54. News.com.au., "Dick Smith is Offering a $5000 Reward for Anyone Who Can Trace the Origins of the Marree Man." News.com.au website, June 26, 2018.

55. *What on Earth?* Season 3, episode 2, "Mystery in the Outback." Aired November 22, 2016, on Science Channel.

56. Frances Mao, "Marree Man: The Enduring Mystery of a Giant Outback Figure," BBC News website, June 26, 2018.

57. *What on Earth?* "Mystery in the Outback."

58. Jürgen Tampke, *The Germans in Australia* (Port Melbourne: Cambridge University Press, 2006), 45,51.

59. *What on Earth?* "Mystery in the Outback."

60. Sian Powell, "Marree Man Refuses to Divulge His Secret," *The Australian* website, June 25, 2018.

61. Powell, "Marree Man."

62. Michael Schirber, "Attempts to Contact Aliens Date Back More Than 150 Years," Space.com website, January 29, 2009.

63. Valerie J. Fletcher, *Isamu Noguchi Master Sculptor* (Washington, DC: Smithsonian Institution, 2005), 171.

64. Douglas A. Vakoch, *Archaeology, Anthropology, and Interstellar Communication.* The NASA History Series (Createspace Independent Publishing Platform, 2014), xxi–xxix.

65. Sarah Cascone, "NASA Suggests Aliens May Be Behind Ancient Rock Art," Art World website, May 24, 2014.

66. John E. Brandenburg, *Death on Mars: The Discovery of a Planetary Nuclear Massacre* (Kempton, IL: Adventures Unlimited, 2015), 257.

67. Brandenburg, *Death on Mars*, 60–64.

68. Brandenburg, *Death on Mars*, 60–64.

69. Josh Hrala, "Stephen Hawking Warns Us to Stop Reaching Out to Aliens before It's Too Late," Science Alert website, November 4, 2016.

1. The Sagan Pyramid

1. Asif A. Siddiqi, "Deep Space Chronicle: A Chronology of Deep Space and Planetary Probes, 1958–2000," *Monographs in Aerospace History*, no. 24 (June 2002): 89.

2. R. J. Parks, "Mariner 9 and the Exploration of Mars," *Astronautical Research 1972* (1973): 149–162.

3. Elizabeth Howell, "Mariner 9: First Spacecraft to Orbit Mars," Space.com website, November 8, 2018.

4. Mack Gipson and Victor K. Ablordeppey, "Pyramidal Structures on Mars," *Icarus* 22, no. 2 (June 1974): 197–204.

5. Carl Sagan, "Christmas Lectures: The Planets: Mars before Viking," Royal Institution, London 1977.

6. Carl Sagan, *Cosmos* (New York: Random House, 1980), 129, 130.

7. USGS, "Elysium Planitia," Astrogeology Science Center, Gazetteer of Planetary Nomenclature, Mars, 2022.

8. Jonathan O'Callaghan, "Signs of Recent Volcanic Eruption on Mars Hint at Habitats for Life," *New York Times* website (Science), November 20, 2020.

9. J. Nussbaumer, "Possible Sea Sediments due to Glaciofluvial Activity in Elysium Planitia, Mars." (paper presented at European Planetary Science Congress, Berlin, Germany, September 2006), 427.

10. Elizabeth Howell, "InSight Lander: Probing the Martian Interior," Space.com website, November 26, 2018.

11. David Sacks, *A Dictionary of the Ancient Greek World* (New York: Oxford University Press US, 1997), 8, 9.

12. Dr. David R. Williams, "Viking Mission to Mars," NASA website (Planetary/Viking), April 12, 2018.

13. Kevin Wilcox, "Ambitious mission required extensive technical innovation," *This Month in NASA History* (blog), August 8, 2022.

14. Carl Sagan and Ann Druyan, *The Demon-Haunted World: Science as a Candle in the Dark* (Random House Publishing Group, 1997), 51, 52.

15. European Space Agency, "Launch Phase," Mars Express website.

16. Alfred McEwen, "Instruments," NASA website (Mars Reconnaissance Orbiter).

17. Personal communication with James S. Miller, August 7, 2020. Miller is an image analyst and member of The Cydonia Institute. Measurements were made utilizing the ancillary data for MRO HiRISE CTX image P11_005219_1961_XN_16N198W (2007), provided by The University of Arizona: Lunar and Planetary Laboratory.

18. Arizona State University School of Earth and Space Exploration, "About THEMIS & the Mars Odyssey mission." Mars Space Flight Facility, Arizona State University, Mars Odyssey THEMIS website, 2002.

19. Personal email communication with Philip Christensen by James S. Miller, July 18, 2002. Christensen is also the Regents Professor of geological sciences and a Professor in the School of Earth and Space Exploration at Arizona State University.

20. Personal communication with James S. Miller, August 7, 2020. Miller is an image analyst and member of The Cydonia Institute. Measurements were made utilizing the ancillary data for MRO HiRISE CTX image P13_006142_1964_XN_16N198W (2007), provided by The University of Arizona: Lunar and Planetary Laboratory.

21. Scientific American, "The Formation of Mountains by Water, the Influence of Erosion by Water in Modeling the Landscape," *Scientific American Supplement* LXXIL, No. 1859 (New York, August 19, 1911), 124.

22. Emmet Gowin and Robert Adams, *The Nevada Test Site* (Princeton: Princeton University Press, 2019), 148–157.

23. Lawrence Livermore National Laboratory, Weapons and Complex Integration, "Big Facility, Explosives Experimental Facility," LLNL website.

24. Personal communication with James S. Miller, August 7, 2020. Measurements were made utilizing the ancillary data for MRO CTX image B20_017574_1965_XN_16N198W (2010), provided by The University of Arizona: Lunar and Planetary Laboratory.

25. Josh Rubin, "2016 TED Prize Winner: Dr Sarah Parcak's Crowdsourcing Space Archaeology," Cool Hunting website, February 17, 2016.

26. Personal communication with James S. Miller, August 7, 2020. Measurements were made utilizing the ancillary data for MRO HiRISE CTX image G01_018708_1959_XN_15N198W (2010), provided by The University of Arizona: Lunar and Planetary Laboratory.

27. Personal communication with James S. Miller, August 7, 2020. Measurements were made utilizing the ancillary data for MRO HiRISE CTX image D06_029600_1968_XN_16N198W (2012), provided by The University of Arizona: Lunar and Planetary Laboratory.

28. Kendall D. Gott, *Where the South Lost the War: An Analysis of the Fort Henry—Fort Donelson Campaign, February 1862* (Mechanicsburg, PA: Stackpole Books, 2003), 73.

29. Personal communication with James S. Miller, August 7, 2020. Measurements were made utilizing the ancillary data for MRO HiRISE CTX image P03_002318_1961_XN_16N198W (2007), provided by The University of Arizona: Lunar and Planetary Laboratory.

30. William Stukeley, *Abury, a Temple of the British Druids, with Some Others, Described, Volume 2* (London: 1743), 36.

31. William Stukeley, "Avebury a Present from the Past: The Longstones," Avebury UK website.

32. Steve Marshall, "Exploring Avebury: The Essential Guide," *The Antiquaries Journal* 97 (September 2017): 314–315.

33. F. K. Annable and D. D. A. Simpson, *A Guide Catalogue of the Neolithic and Bronze Age Collections in Devizes Museum* (Devizes, Wiltshire: Wiltshire Archaeological and Natural History Society, 1964), 66, 122.

34. I. F. Smith and J. G. Evans, "Excavation of Two Long Barrows in North Wiltshire," *Antiquity* 42, no. 166 (June 1, 1968): 138.

35. William J. Toman, "Lizard Effigy Mound, 500–1000 A.D.," The Historical Marker Database website, July 16, 2010.

36. James P. Scherz and Buck Trawicky, "Survey Report Hudson Park Mound Group—Dane County Madison Wisconsin," Ancient America website, March 24, 2017.

37. Personal communication with Michael Dale, January 2020. Dale is a geologist and a member of The Cydonia Institute.

38. Personal communication with Michael Dale, January 2020.

39. Richard F. Townsend, *Hero, Hawk, and Open Hand: American Indian Art of the Ancient Midwest and South* (Chicago: Art Institute of Chicago in association with Yale University Press, 2004), 17.

40. Mike Skele, "The Great Knob: Interpretations of Monks Mound," *Studies in Illinois Archaeology*, no. 4 (Springfield, IL: Illinois Historic Preservation Agency, 1988), 1–3.

41. William M. Denevan. "The Pristine Myth: The Landscape of the Americas in 1492," *Annals of the Association of American Geographers* 82, no. 3 (September 1992): 369–385.

42. Biloine W. Young and Melvin Leo Fowler, *Cahokia: The Great Native American Metropolis* (Urbana: University of Illinois Press, 2000), 288.

43. Ephraim G. Squier and Edwin H. Davis, *Contributions to Knowledge Vol 1.* (Washington, DC: Smithsonian Institution, 1848).

44. Tristram R. Kidder, "Excavations at the Jordan Site (16MO1), Morehouse Parish, Louisiana." *Southeastern Archaeology* 11, no. 2 (Winter 1992): 109–131.

45. John H Blitz, *Moundville* (Tuscaloosa: University of Alabama Press, 2008), 3.

46. Sagan, *Cosmos*, 92.

2. Darkness on the Edge of Forever I

1. Mean Green, "Another Interesting Area WITH Possible Structures/Ruins ;-)" Facebook, April 26, 2018.

2. NASA, "Nepenthes Mensae," Jet Propulsion Laboratory Image website, April 11, 2012.

3. Miguel Ángel De Pablo and Andrea Pacifici, "Geomorphological Evidence of Water Level Changes in Nepenthes Mensae, Mars." *Icarus* 196, no. 2 (August 2008): 667–671.

4. USGS, "Nepenthes Mensae," Astrogeology Research Center, Gazetteer of Planetary Nomenclature, Mars, 2020.

5. Homer, *The Odyssey with an English Translation by A.T. Murray, Ph.D., Volume 1 & 2.* (Cambridge, MA: Harvard University Press; London: William Heinemann, Ltd., 1919), 219–221.

6. Stephen C. Spiteri, "Illustrated Glossary of Terms Used in Military Architecture Terms." *Arx: International Journal of Military Architecture and Fortification* (2010): 635–650.

7. Gott, *Where the South Lost the War*, 73.

8. John E. Kleber, *The Kentucky Encyclopedia* (Lexington: University Press of Kentucky, 2014), 534.

9. A. Valenciano et al., "The Role of Water on the Evolution of the Nepenthes Mensa Region of Mars," 40th Lunar and Planetary Science Conference, March 2009, 1052.

10. Mars Viewer, MRO HiRISE CTX, G21_026542_1894_XN_09N235W, "Nepenthes Planum," Arizona State University Mars Space Flight Facility website (MRO Context Camera), March 25, 2012.

11. Minh Tuan, "European Investors Eye US $1 Billion Logistics Project in Vietnam," *The Saigon Times*, September 20, 2020.

12. Irene Guzman, "The (Triangular) Office Building Designed by Jo Coenen and Archisquare in Parma, Abitare," Abitare website, June 17, 2015.

13. Nancy B. Simmons, *Mammal Species of the World: A Taxonomic and Geographic Reference*, 3rd ed. (Baltimore: Johns Hopkins University Press, 2005), 312–529.

14. David McDonald, *The Encyclopedia of Mammals* (Oxford University Press, 2010), 466.

15. Hans-Ulrich Schnitzler et al., "Fishing and Echolocation Behavior of the Greater Bulldog Bat, *Noctilio leporinus*, in the Field," *Behavioral Ecology and Sociobiology* 35, no. 5 (November 1994): 327–345.

16. Bill Holm, *Northwest Coast Indian Art an Analysis of Form* (Seattle: University of Washington Press, 1965), 37.

17. Clint Leung, *An Overview of Pacific Northwest Native Indian Art, Exquisite Pacific Northwest Native Indian & Inuit Art* (Free Sprit Gallery, 2006), 18.

18. Jean Chevalier and Alain Gheerbrant, *A Dictionary of Symbols* (London: Penguin Books, 1996), 747.

19. Jason Yaeger, "Untangling the Ties That Bind: The City, the Countryside, and the Nature of Maya Urbanism at Xunantunich, Belize," in *The Social Construction of Ancient Cities*. Edited by Monica L. Smith (Washington, DC: Smithsonian Institution Press, 2003), 121–155.

20. Alfred Newton, *A Dictionary of Birds* (Harvard University, A. and C. Black, 1893): 160.

21. Susumu Ohno, *Sex Chromosomes and Sex-Linked Genes, Volume 1, Monographs on Endocrinology* (Berlin: Springer Science & Business Media, 2013), 146.

22. Wendell M. Levi, *The Pigeon* (Columbia, SC: R. L. Bryan Company, 1941), 38, 73.

23. Allevamento Poggio Di Ponte Breeding Center, "Seldschuk Dove," Allevamento Poggio Di Ponte Breeding Center website, 2022.

24. Roz Payne, "Dove in Pentagon," Sixties Archive website.

3. Darkness on the Edge of Forever II

1. William R. Saunders, "Mean City: Darkness on the Edge of Forever," The Cydonia Institute Discussion Board, May 18, 2018.

2. Personal conversation with James S. Miller, March 22, 2023.

3. W. G. McLachlan, "The Mounds of Lake Waubesa Region," in *The Wisconsin Archeologist* 11–13, Wisconsin Natural History Society, *University of Chicago Wisconsin Archeological Society*, 12, no. 4 (January 1914): 111–113.

4. John Noble Wilford, "Mapping Ancient Civilization, in a Matter of Days," *New York Times*, May 10, 2010.

4. Elongated Hexagonal Mound

1. Personal communication with Gary Leggiere, October 27, 2011.

2. USGS, "Nepenthes Mensae."

3. Personal email communication with geomorphologist William R. Saunders and geologist Michael Dale, 2011.

4. Personal email communication with geomorphologist William R. Saunders and geologist Michael Dale, 2011.

5. NASA, "Mars Global Surveyor."

6. Kavita Gangal et al., "The Near-Eastern Roots of the Neolithic in South Asia," *PloS One* 9, no. 5 (2014): 1.

7. Jesse Casana and Jackson Cothren, "The CORONA Atlas Project: Orthorectification of CORONA Satellite Imagery and Regional-Scale Archaeological Exploration in the Near East," *Mapping Archaeological Landscapes from Space, SpringerBriefs in Archaeology*, Vol 5. (New York: Springer, 2013), 38.

8. Douglas C. Comer and Michael J. Harrower, *Mapping Archaeological Landscapes from Space* (New York: Springer, 2013), 38.

5. The Keyhole

1. McEwen, Alfred, MRO HiRISE ESP_020794_1860, "Exclamation Mark on Mars," University of Arizona HiRISE website, January 2, 2011.

2. McEwen, "Exclamation Mark on Mars."

3. Personal email communication from Greg Orme, July 17, 2013.

4. George J. Haas, "Keyhole—Exclamation Mark on Mars," The Cydonia Institute Discussion Board, July 17, 2013.

5. USGS, "Libya Montes," Astrogeology Science Center, Gazetteer of Planetary Nomenclature, Mars, 2022.

6. Janice L. Bishop et al., "Mineralogy and Morphology of Geologic Units at Libya Montes, Mars: Ancient Aqueously Derived Outcrops, Mafic Flows, Fluvial Features, and Impacts," *Journal of Geophysical Research: Planets* 118, no. 3 (March 31, 2013): 347–576.

7. Jet Propulsion Laboratory, "A Regional View of the Libya Montes," NASA's Mars Images website, May 22, 2000.

8. Alfred S. McEwen, "Faculty," The University of Arizona, Lunar and Planetary Laboratory website.

9. McEwen, "Exclamation Mark on Mars."

10. Personal email communication from Greg Orme, July 17, 2013.

11. William R. Saunders et al., "A Wedge and Dome Formation Set within the Flat Plains of Libya Montes," *Journal of Space Exploration* 4, no. 3 (November 17, 2016): 1–14.

12. Saunders et al., "A Wedge and Dome Formation."

13. David L. Kennedy, "Desktop Archeology," *Aramco World* 60, no. 4 (July/August 2009).

14. Rebecca Kessler, "Thousands of Tombs in Saudi Desert Spotted from Space," Live Science website, February 15, 2011.

15. Amelia Carolina, Sparavigna, "A Prehistoric Solar Observatory in the Middle of Sahara Desert," *PHILICA*, Article 428 (October 22, 2014), 1–7.

16. Wendy Doniger, ed., *Merriam-Webster's Encyclopedia of World Religions* (Springfield, MA: Merriam-Webster, 1999), 179.

17. Stephen Denison Peet, *State Historical Society of Wisconsin, Collections—State Historical Society of Wisconsin* (The University of California, General Books LLC, 1882), 49.

18. Alma M. Reed, *The Ancient Past of Mexico* (New York: Crown Publishers Inc., 1966), 12.

19. Vik Muniz, "Pictures of Earthwork," *Nichido Contemporary Art*, 2009.

20. Aileen Kawagoe, "4. Towering Tumuli of the Kofun Era," *Heritage of Japan* (blog), 2013.

21. Robert Schoch, *Ancient Aliens*, season 5, episode 6, "Secrets of the Tombs," aired January 25, 2013, on History Channel.

6. The Martian Atlantis I

1. Greg Orme, "The Kings Valley, Mars, Why We Must Go to Mars," Facebook, July 17, 2019.

2. Javed Raza, "A 'City' on MARS in the Atlantis Chaos Region," Facebook, July 24, 2019.

3. Neville Thompson, "Javed Raza's-Atlantis Chassis-ESP 019103 1460," *Gigapan* web-site, August 7, 2019.

4. Personal communication with image analyst James S. Miller, July 27, 2022.

5. European Space Agency, "Chaos in Atlantis Basin," ESA website: Mars Express Space Science image of the week, October 27, 2014.

6. USGS, "Atlantis Chaos," Astrogeology Research Center, Gazetteer of Planetary Nomenclature, Mars, 2022.

7. European Space Agency, "Chaos in Atlantis basin."

8. USGS, "Atlantis Chaos."

9. Robert Graves, *The Greek Myths* (London, UK: Moyer/Bell, 1994), 33.

10. Marilu Waybourn, *Images of America: Aztec* (Charleston, SC: Arcadia Publishing 2011), 7.

11. Newsgram Desk, "UNESCO Declares Al 'Ula as Saudi's First World Heritage Site: The 2,000-Year-Old Town Is Made of Mud and Stone," June 14, 2016.

12. Pedro Sueldo Nava, *A Walking Tour of Machupicchu: Editorial de Cultura Andina.* (University of Texas, 1976), 9–10.

13. Personal communication with image analyst James S. Miller, July 27, 2022.

14. Javed Raza, "Archaeology on Mars Series . . . Discovery of an Earth Like 'City' & Structures," Facebook, July 27, 2019.

15. John M. Fritz and George Mitchell, *Hampi Vijayanagara* (Jaico Pub House, 2015), 58–65.

16. Tafline Laylin, "Agora Tower: Twisting Skyscraper Wrapped with Vertical Gardens Breaks Ground in Taipei," Inhabitat website, March 10, 2013.

17. Personal communication with image analyst James S. Miller, July 27, 2022.

18. Malinga Amarasinghe, *The Ruins of Polonnaruwa* (in Sinhala) (S. Godage & Brothers, 1998), 75–77.

19. Personal communication with James S. Miller, July 28, 2022.

20. Chester H. Liebs, "Inventory—Nomination form Old Red Mill," National Register Historic Places website, May 8, 1972

21. Personal communication with geomorphologist William R. Saunders, August 8, 2019.

22. D. E. Leaman, "The Nature of Jointing, Tasman Peninsula, Southern Tasmania." *Papers and Proceedings of the Royal Society of Tasmania* 133, no. 1 (1999): 65–76.

23. Derek Ford and Paul Williams, *Karst Hydrogeology and Geomorphology* (Chichester, England: John Wiley & Sons Ltd., 2007), 1.

24. R. Mckeever et al., "Basalt Weathering in a Temperate Marine Environment," *Developments in Soil Science* 19 (1990), 519–523.

25. Ennio Piccaluga, "The Cities of Mars," *Epiccaluga Wordpress* (blog), October 12, 2019.

26. National Park Service, "How Big Are the Heads?" National Park Service Website for Mount Rushmore National Memorial, South Dakota, April 2, 2020.

27. Peter Herrle and Erik Wegerhoff, eds. "Longstones Barrow" In *Architecture and Identity*, Volume 9, Habitat—International LIT (Verlag Münster, 2008), 40–44.

28. Michael D. Coe and David Grove. *The Olmec and Their Neighbors: Essays in Memory of Matthew W. Stirling* (Washington, DC: Dumbarton Oaks, 1981), 318.

29. Mark Miller Graham, *Jade in Ancient Costa Rica* (New York: The Metropolitan Museum of Art, 1998), 53.

30. Marlin Calvo Mora, *Gold, Jade, Forests: Costa Rica* (University of Washington Press, 1995), 51.

7. The Martian Atlantis II

1. USGS, "Atlantis Chaos."

2. Harry Seidler, "Riverside Center," Harry Seidler & Associates website (Offices and Public Buildings), 1986.

3. Emile Mayer, "Brialmont," *Revue d'histoire modern* 2, no. 9 (1927): 2023177–190.

4. Clayton Donnell, *Breaking the Fortress Line 1914* (Barnsley, South Yorkshire, UK: Pen and Sword, 2013), 12.

5. Personal email communication with geomorphologist William R. Saunders, November 30, 2022.

6. Carl Bovill, *Fractal Geometry in Architecture and Design* (Boston: Birkauser, 1996), 3.

7. Michelle N. Stevens, "Archaeological Preservation and Environmental Conservation in Arizona's Cienega Valle," *Archaeology Southwest* 15, no. 4 (Fall 2011): 11.

8. Higyou, "Several Small Modern Buildings following an Arrow-Shaped One, 3D, Isolated," Image ID: MH45F5, Alamy website, May 1, 2018.

9. Naglaa A. Megahed, "Origami Folding and Its Potential for Architecture Students," *The Design Journal: An International Journal for All Aspects of Design* 20, no. 2 (January 16, 2017): 279–297.

10. Adam Phillips, "Interview with Ground Zero Architect Daniel Libeskind—2003-03-23," *Voa News* website, October 26, 2009.

11. Michael J. Craig, "Geometrical Features on Mars, in 'Atlantis Chaos," Secret Mars, The Mars Archaeology Archive website.

12. Measurements provided by image analyst James S. Miller, September 4, 2019.

13. Personal conversation with image analyst James S. Miller, September 20, 2019.
14. Personal email communication with geomorphologist William R. Saunders, November 30, 2022.

8. The Martian Atlantis III

1. European Space Agency, "Chaos in Atlantis basin."
2. Chevalier and Gheerbrant, *Dictionary of Symbols*, 710–711.
3. Marten Kuilman, "Quadralectic Architecture—A Panoramic Review," *Quadralectic Architecture* (blog).
4. Guzman, "The (Triangular) Office Building by Jo Coenen."
5. Pozos, *The Face on Mars,* 72.
6. Mars Viewer, Mars Odyssey, THEMIS, V13700005, Arizona State University Mars Space Flight Facility website (THEMIS), January 15, 2005.
7. Chevalier and Gheerbrant, *Directory of Symbols*, 189.
8. Paul Jankowski, *Verdun: The Longest Battle of the Great War* (Oxford: Oxford University Press, 2014), 111.
9. Brian Yare, "The Middle Kingdom Egyptian Fortresses in Nubia," Yare website, January 28, 2001.
10. Craig, "Geometrical Features on Mars, in 'Atlantis Chaos.'"
11. Measurements provided by image analyst James S. Miller. September 3, 2019.
12. Personal conversation with image analyst James S. Miller, September 20, 2019.

9. Parrotopia I

1. Mars Viewer, MOC M1402185. "S-Facing Slope on Massif in W Argyre Rim Region," Arizona State University Mars Space Flight Facility website (Mars Orbiter Camera), April 30, 2000.
2. Wil Faust, "Parrotopia Update No.1: Was the City a Port?" Parrotopia—Anomaly Hunters Roundtable Study website, January 10, 2003.
3. R. DeRosa, "Parrotopia, ID 97522, Layered Knobs and Rectilinear Ridges in Nereidum Montes," MRO HiRISE website, March 9, 2014.
4. DeRosa, "Parrotopia."
5. USGS, "Argyre Planitia," Astrogeology Science Center, Gazetteer of Planetary Nomenclature, Mars, 2022.
6. Walter S. Kiefer et al., "The Red Planet: A Survey of Mars," 2nd ed. Slide 15. Lunar and Planetary Institute website, 2011.
7. Harold Hiesinger and James W. Head, III, "Topography and Morphology of the Argyre Basin, Mars: Implications for Its Geologic and Hydrologic

History," *Planetary and Space Science* 50, no. 10–11 (August–September 2002): 939–981.

8. John Bostock and Henry Thomas Riley, *The Natural History of Pliny, Volume 2* (London: H. G. Bohn, 1855), National Library of Medicine website.

9. Saunders, William R., George J. Haas, and James Miller, "A Persistent Avian Formation on a South-Facing Slope, along the Northwest Rim of the Argyre Basin of Mars," *Journal of Scientific Exploration* 36, no. 2 (Summer 2022): 335.

10. Dale et al., "Avian Formation on a South-Facing Slope," 524.

11. Ronald Greeley et al., "Martian Aolian Processes, Sediments, and Features," Mars (1992): 730–766.

12. Hiesinger and Head, "Topography and morphology of the Argyre Basin, Mars," 939–981.

13. Dale et al., "Avian Formation on a South-Facing Slope," 524.

14. Dale et al., "Avian Formation on a South-Facing Slope," 525.

15. Dale et al., "Avian Formation on a South-Facing Slope," 525, 526.

16. B. Grzimek, *Grzimek's Animal Life Encyclopedia*, Volume 9, Birds II, 2nd ed. (Thomson/Gale, 2003), 275.

17. Dale et al., "Avian Formation on a South-Facing Slope," 530.

18. Dale et al., "Avian Formation on a South-Facing Slope," 530.

19. Dale et al., "Avian Formation on a South-Facing Slope," 532, 533.

20. Saunders et al., "A Persistent Avian Formation on a South-Facing Slope," 337.

21. David Hurst Thomas, *Exploring Ancient Native America: An Archaeological Guide* (New York: Macmillan, 1994), 135.

22. Personal email communication with Dr. Susan Orosz, November 21, 2006.

23. Max E. White, *The Archaeology and History of the Native Georgia Tribes* (Gainesville, FL: University Press of Florida, 2002), 52.

24. Maria Longhena and Walter Alva, *The Incas and Other Andean Civilizations* (San Diego: Thunder Bay Press, 1999), 198.

25. Malin Space Science Systems, "West Argyre, MGS MOC Release No. M0C2-1191," Mars Global Surveyor, Mars Orbiter Camera website, August 22, 2005.

26. Phillips, "Earthlings Look for Signs in New Photos of Mars," 1, A10.

10. Parrotopia II

1. Faust, Faust, "Parrotopia Update No.1."

2. Mars Viewer, MOC M1402185.

3. Personal email communication with image analyst James S. Miller, November 10, 2022.

4. Mike Wicks, "What Does Accent Inns and the Royal Canadian Navy Have in Common?" *Accent Inns* (blog), 2017.

5. Felipe Solis and Ted Leyenaar, *Mexico: Journey to the Land of the Gods* (Amsterdam, Netherlands: Lund Humphries, 2002), 184.

6. Tafline Laylin, "Luis de Garrido Unveils Incredible Glass-Domed Eco House Shaped Like the Eye of Horus," Inhabitat website, September 17, 2011.

7. Karl Andreas Taube, *The Major Gods of Ancient Yucatan* (Washington, DC: Dumbarton Oaks Research Library, 1992), 133.

8. Alfred Marston Tozzer and Glover Morrill Allen, "Animal Figs in the Maya Codices," Volume IV, no. 3 of *Papers of the Peabody Museum of American Archaeology and Ethnology* (Cambridge, MA: Harvard University/Harvard Museum, 1910), 345.

9. Sophie D. Coe and Michael D. Coe, *The True History of Chocolate* (New York: Thames and Hudson, 1996), 40, 41.

10. Carlos Pallán Gayol, "The Many Faces of Chaahk: Exploring the Role of a Complex and Fluid Entity within Myth, Religion and Politics," In "The Maya and Their Sacred Narratives: Text and Context in Maya Mythologies," *Acta Mesoamericana* 20 (2009): 31.

11. Mary Miller and Karl Taube, *An Illustrated Dictionary of The Gods and Symbols of Ancient Mexico and the Maya* (New York: Thames and Hudson, 2015), 166.

12. Candis M. Keener, "The Baby Jaguar Series: A Comparative Analysis" (MA thesis, Kent State University, 2009), 78.

13. Fredrick W. Lange, *Pre-Columbian Jade: New Geological and Cultural Interpolations* (Salt Lake City: University of Utah Press, 1993), 265.

14. Edwin Barnhart, "The First Twenty-Three Pages of the Dresden Codex: The Divination Pages" (Thesis, Graduate School of the University of Texas, Austin, May 1996), 19.

15. Taube, *The Major Gods of Ancient Yucatan*, 101.

16. Carol Dommermuth-Costa, *Nikola Tesla: A Spark of Genius* (Twenty-First Century Books, 1994), 75.

17. William S. Weed, "Tower of Power: What Happened to Engineer Greg Leyh When He Climbed to the Top of a Giant Electrical Transformer That Throws Lightning Bolts into the Sky?—Physical—Electrum, World's Largest Tesla Coil—Cover Story," Lightning on Demand website (Current Science), January 2, 2004, 4, 5.

18. Taube, *The Major Gods of Ancient Yucatan*, 75.

19. Javier Urcid, *Zapotec Writing: Knowledge, Power, and Memory in Ancient Oaxaca,* (Department of Anthropology, Brandeis University, May 2005), 139.

20. Mary Ellen Miller, "A Re-Examination of the Mesoamerican Chacmool," *The Art Bulletin* 67, no. 1 (March 1985): 7–17.

21. Personal email communication with image analyst, James S. Miller, November 17, 2022.

22. Brian S. Bauer, *Ancient Cuzco: Heartland of the Inca* (Austin, TX.: University of Texas Press, 2010), 3.

23. Andrew Rogers, "Lion's Paw," Andrewrogers.org website (Land Art/Kenya), 2010.

24. Hartley's Safaris, "Sculpture in Chyulu Hills Kenya Seen from Space," Hartley's Safaris website, "Latest News," March 11, 2012.

25. Claude-Francois Baudez, *Maya Sculpture of Copán: The Iconography* (Norman: University of Oklahoma Press, 2015), 27.

26. Linda Schele and David Freidel, *A Forest of Kings: The Untold Story of the Ancient Maya* (New York: Quill, 1990), 411.

27. Lynn V. Foster, *Handbook to Life in the Ancient Maya World* (Oxford: Oxford University Press, 2005), 185.

28. Wil Faust, "Parrotopia Update No.1: Was the City a Port?" Parrotopia—Anomaly Hunters Roundtable Study website, January 10, 2003.

29. Jim Miller, "The Anomaly Hunters Roundtable Study: Parrotopia, the Parrot and City Complex," Anomaly Hunters website, January 18, 2003.

30. Robert M. Carmack, *The Quiché Mayas of Utatlán: The Evolution of a Highland Guatemala Kingdom* (Norman: University of Oklahoma Press, 1981), 214.

31. Thomas F. Babcock, *Utatlán, The Constituted Community of the K'iche' Maya of Q'umarkaj* (Boulder: University Press of Colorado, 2012), 5.

32. Joyce Kelly, *An Archaeological Guide to North Central America: Belize, Guatemala, Honduras, and El Salvador* (Norman: University of Oklahoma Press, 1996), 200.

11. The Hero Twins and the Turtle of Creation

1. Karen Bassie-Sweet, "Corn Deities and the Complementary Male/Female Principle," *La Tercera Mesa Redonda de Palenque* (July 1999, Revised September 2000): 5.

2. Karen Bassie-Sweet, "Corn Deities," 5.

3. Susan Milbrath, *Star Gods of the Maya: Astronomy in Art, Folklore, and Calendars* (Austin: University of Texas Press, 1999), 150.

4. Milbrath, *Star Gods of the Maya: Astronomy in Art*, 151.

5. Miller and Taube, *An Illustrated Dictionary*, 137.

6. Dennis Tedlock, *Popol Vuh: the Definitive Edition of the Mayan Book of the Dawn of Life and the Glories of Gods and Kings* (New York: Touchstone, 1986), 90–94.

7. Mark Carwardine, *Animal Records* (New York: Sterling Publishing Company, Inc.), 174.

8. Chevalier and Gheerbrant, *Dictionary of Symbols*, 1016.

9. Miller and Taube, *An Illustrated Dictionary*, 175.

10. Marc Zender, "Teasing the Turtle from Its Shell: AHK and MAHK in Maya Writing." *PARI Journal* VI, no. 3 (Winter 2006): 9.

11. David Freidel et al., *Maya Cosmos: Three Thousand Years on the Shaman's Path* (New York: Quill, 1995), 82.

12. Linda Schele and Peter Mathews, *The Code of Kings: The Language of Seven Sacred Maya Temples and Tombs* (New York: Touchstone Books/Simon & Schuster, 1999), 37.

13. Zender, "Teasing the Turtle from Its Shell," 8.

14. David Stuart, *The Inscriptions from Temple XIX at Palenque* (San Francisco: Pre-Columbian Art Research Institute, 2005), 23.

15. Freidel et al., *Maya Cosmos*, 79.

16. Daniel J. Field et al., "Toward Consilience in Reptile Phylogeny: miRNAs Support an Archosaur, Not Lepidosaur, Affinity for Turtles," *Evolution & Development* 16, no. 4 (July/August 2014): 189–196.

17. Stuart, *The Inscriptions from Temple XIX at Palenque*, 151.

12. The Anunnaki

1. Zecharia Sitchin, *The 12th Planet: Book I of The Earth Chronicles* (New York: Avon Books, 1976), 328.

2. Sitchin, *The 12th Planet*, vii.

3. Sitchin, *The 12th Planet*, 230.

4. Miller and Taube, *An Illustrated Dictionary*, 167, 168.

5. Brockhampton Press, *Dictionary of Classical Mythology* (Leicester, UK: Brockhampton Press LTD, 1995), 117.

6. Michael D. Coe, *Breaking the Maya Code* (New York: Thames and Hudson, 1992), 133.

7. Sitchin, *The 12th Planet*, 214.

8. Sitchin, *The Lost Book of Enki*, 29–38.

9. Sitchin, *The 12th Planet*, 291.

10. Sitchin, *The 12th Planet*, 253.

11. Zecharia Sitchin, *Genesis Revisited: Is Modern Science Catching up with Ancient Knowledge?* (New York: Avon Books, 1990), 183.

12. Zecharia Sitchin, *The Lost Book of Enki: Memoirs and Prophecies of an Extraterrestrial God* (Rochester, VT: Bear & Company, 2002), 104.

13. Sitchin, *The 12th Planet*, 328.

14. Sitchin, *The 12th Planet*, 327.

15. Sitchin, *The 12th Planet*, 336–361.

16. Sitchin, *The 12th Planet*, 341.

17. Sitchin, *The 12th Planet*, 350.

18. Sitchin, *The 12th Planet*, 350.

19. Sitchin, *The 12th Planet*, 351.

20. Cottie Burland, et al., *Mythology of the Americas* (London: Hamlyn, 1970), 155.

21. Sitchin, *Genesis Revisited*, 185–189.

22. Sitchin, *The Lost Book of Enki*, 178–182.

23. Sitchin, *The 12th Planet*, 381.

24. Zecharia Sitchin, *The Wars of Gods and Men: Book III of The Earth Chronicles* (New York: Avon Books, 1985), 346.

25. Ross, "First City in the New World?" 60.

26. Sitchin, *The Wars of Gods and Men*, 35.

27. Sitchin, *The Wars of Gods and Men*, 35.

28. Richard A. Diehl, *The Olmecs: America's First Civilization* (London: Thames and Hudson, 2004), 12.

29. Zecharia Sitchin, *The Lost Realms: Book IV of The Earth Chronicles* (New York: Avon Books, 1990), 269.

30. Sitchin, *The Lost Realms*, 269.

31. Adrian Recinos, *Popol Vuh: The Sacred Book of the Ancient Quiché Maya*, Civilization of the American Indian Series. Translated by Delia Goetz and Sylvanus Griswold Morley (Norman: University of Oklahoma Press, 1950), 80.

13. The Maya "Star-War"

1. Simon Martin and Nikolai Grube, *Chronicle of the Maya Kings and Queens: Deciphering the Dynasties of the Ancient Maya* (London: Thames and Hudson, 2000), 69.

2. Martin and Grube, *Chronicle of the Maya Kings and Queens*, 69, 72.

3. Spencer T. Mitchell, "Visual Communications of Power: The Iconography of the Classic Maya Naranjo-Sa'al Polity" (MA thesis, Texas Tech University, December 2016), 1.

4. Mitchell, "Visual Communications of Power," 1.

5. Martin and Grube, *Chronicle of the Maya Kings and Queens*, 70.

6. Sylvanus G. Morley, "The Inscriptions of Naranjo, Northern Guatemala," *American Anthropologist* 11, no. 4 (October–December 1909): 543–562.

7. Mitchell, "Visual Communications of Power," 9.

8. Mitchell, "Visual Communications of Power," 9.

9. Mitchell, "Visual Communications of Power," 10.

10. John Montgomery, *Dictionary of Maya Hieroglyphics* (New York: Hippocrene Books, Inc., 2002), 161.

11. Montgomery, *Dictionary of Maya Hieroglyphics*, 94.

12. Montgomery, *Dictionary of Maya Hieroglyphics*, 212.

13. Montgomery, *Dictionary of Maya Hieroglyphics*, 80.

14. Michael D. Coe, *The Olmec World: Ritual and Rulership* (Princeton: The Art Museum, Princeton University, 1996), 36.

15. Martin and Grube, *Chronicle of the Maya Kings and Queens*, 79.

16. J. Eric S. Thompson, *Maya Hieroglyphic Writing* (Norman: University of Oklahoma Press, 1971), 148.

17. Jorge L. Orejel, "The 'Axe/Comb' Glyph as Ch'ak," *Reports on Ancient Maya Writing*, 31. (Washington, DC: Center for Maya Research, 1990), 7.

18. Orejel, "The 'Axe/Comb' Glyph as Ch'ak," 7.

19. Milbrath, *Star Gods of the Maya: Astronomy in Art*, 242.

20. Coe, *Breaking the Maya Code*, 133.

21. Miller and Taube, *An Illustrated Dictionary*, 168; Also see Montgomery, *Dictionary of Maya Hieroglyphics*, 217.

22. Michael J. Grofe, "The Recipe for Rebirth: Cacao as Fish in the Mythology and Symbolism of the Ancient Maya," Department of Native American Studies, University of California at Davis (September 23, 2007), 1.

23. Miller and Taube, *An Illustrated Dictionary*, 68.

24. Miller and Taube, *An Illustrated Dictionary*, 68.

25. Schele and Matthews, *The Code of Kings*, 214.

26. Miller and Taube, *An Illustrated Dictionary*, 174.

27. Mark Pitts, *Maya Numbers and the Maya Calendar: A Non-Technical Introduction to Maya Glyphs—Maya Book 2*. Foundation for the Advancement of Mesoamerican Studies (FAMSI) website, 2009, 8.

28. Montgomery, *Dictionary of Maya Hieroglyphics*, 169.

29. Martin and Grube, *Chronicle of the Maya Kings and Queens*, 74.

30. Schele and Freidel, *A Forest of Kings*, 185.

31. Martin and Grube, *Chronicle of the Maya Kings and Queens*, 78.

32. Martin and Grube, *Chronicle of the Maya Kings and Queens*, 79.

33. Schele and Freidel, *A Forest of Kings*, 147.

34. Montgomery, *Dictionary of Maya Hieroglyphics*, 89.

35. Montgomery, *Dictionary of Maya Hieroglyphics*, 294.

36. Montgomery, *Dictionary of Maya Hieroglyphics*, 130.

37. Montgomery, *Dictionary of Maya Hieroglyphics*, 180.

38. Orejel, "The 'Axe/Comb' Glyph as Ch'ak," 4.

39. Chevalier and Gheerbrant, *A Dictionary of Symbols*, 921.

40. Edwin Braakhuis, "The Way of All Flesh: Sexual Implications of the Mayan Hunt," *Anthropos* 96, no. 2 (2001): 391–409.

41. Montgomery, *Dictionary of Maya Hieroglyphics*, 263.

42. Schele and Freidel, *A Forest of Kings*, 371.

43. Schele and Freidel, *A Forest of Kings*, 393.

44. W. G. Lambert, "The God Aššur," *Iraq* 45, no. 1 (Spring 1983): 82–86.

45. John E. Brandenburg, "Evidence of a Massive Thermonuclear Explosion on Mars in the Past: The Cydonian Hypothesis and Fermi's Paradox," *AIAA Space Forum*, Long Beach California, September 16, 2016, 1–56.

46. Brandenburg, *Death on Mars*, 252–254.

47. Linda Schele, "The Universe: Now and Beyond," NASA Administrator's Third Seminar Series, NASA News website, March 13, 1995.

48. Robert Thompson, Jr., "Linda Schele, Pioneer in the Study of Mayans, Dies at 55," *New York Times*, April 22, 1998.

49. Nikolai Grube and Linda Schele, "Kuy, the Owl of Omen and War," *Mexicon* 16, no. 1 (February 1994): 10–17.

50. Arthur G. Miller, *The Codex Nuttall* (New York: Dover, 1975), ixv-xv.

51. Reiko Ishihara et al. "The Water Lily Serpent Stucco Masks at Caracol, Archaeological Investigations in the Eastern Maya Lowlands: Papers of the 2005 Belize Archaeology Symposium, Belize," *Research Reports in Belizean Archaeology* 3 (2006): 212.

52. Verónica Amellali Vázquez López, "Pact and Marriage: Sociopolitical Strategies of the Kanu'l Dynasty and Its Allies during the Late Classic Period," *Contributions in New World Archaeology* 11 (September 2017): 25.

53. Nicholas Matthew Hellmuth, "Structure 5D-73, Burial 196, Tikal, Peten, Guatemala: A Preliminary Report," (BA honors thesis, Foundation for Latin American Anthropological Research, March 31, 1967), i.

54. Martin and Grube, *Chronicle of the Maya Kings and Queens*, 48–49.

55. Martin and Grube, *Chronicle of the Maya Kings and Queens*, 49.

56. Schele and Freidel, *A Forest of Kings*, 90.

57. Freidel et al., *Maya Cosmos*, 147–148.

58. Schele and Freidel, *A Forest of Kings*, 90.

59. Schele and Freidel, *A Forest of Kings*, 373.

60. Charles Phillips, *The Complete Illustrated History of the Aztec & Maya* (China: Hermes House, 2010), 83.

61. Phillips, *The Complete Illustrated History of the Aztec & Maya*, 82.

62. Phillips, *The Complete Illustrated History of the Aztec & Maya*, 85.

63. Nicholas J. Saunders, "Predators of Culture: Jaguar Symbolism and Mesoamerican Elites," *World Archaeology* 26, no. 1 (June 1994): 104–117.

64. Sue Giles and Valerie Harland, "Digitisation of the Adela Breton Collection at Bristol Museum & Art Gallery," ArtUK website, November 13, 2014.

65. Mary Miller, "The Willfulness of Art: The Case of Bonampak," *RES: Anthropology and Aesthetics* 42 (Autumn 2002): 8–23.

66. John Kelly, "Hard-Core 'Star Wars' Fans Dress Up and Troop Around," *Washington Post* website, December 1, 2013.

67. Recinos, *Popol Vuh*, 80.

68. Acts 17:24–28 (King James Version)

Bibliography

Allevamento Poggio Di Ponte Breeding Center. "Seldschuk Dove." Allevamento Poggio Di Ponte Breeding Center website, 2022. Accessed March 23, 2023.

Amarasinghe, Malinga. *The Ruins of Polonnaruwa* (in Sinhala). S. Godage & Brothers, 1998.

Amellali Vázquez López, Verónica. "Pact and Marriage: Sociopolitical Strategies of the Kanu'l Dynasty and Its Allies during the Late Classic Period." *Contributions in New World Archaeology* 11 (September 2017): 9–48.

Anderson, Donald M. *Elements of Design*. New York: Holt, Rinehart & Winston, 1961.

Andrews, Robin George. "The Middle East Is Dotted with Thousands of Puzzling Kite-Shaped Structures." Atlas Obscura website, January 22, 2019. Accessed March 23, 2023.

Angelo, Joseph A. *Encyclopedia of Space and Astronomy*. New York: Infobase Publishing, 2014.

Annable, F. K., and D. D. A. Simpson. *A Guide Catalogue of the Neolithic and Bronze Age Collections in Devizes Museum*. Devizes, Wiltshire: Wiltshire Archaeological and Natural History Society, 1964.

Arizona State University School of Earth and Space Exploration. "About THEMIS & the Mars Odyssey mission." Mars Space Flight Facility, Arizona State University Mars Odyssey THEMIS website, 2002. Accessed March 23, 2023.

Aveni, Anthony F. "Solving the Mystery of the Nasca Lines." *Archaeology* 53, no. 3 (May/June 2000): 26–35.

Babcock, Thomas F. *Utatlán, The Constituted Community of the K'iche' Maya of Q'umarkaj*. Boulder: University Press of Colorado, 2012.

Barnhart, Edwin. "The First Twenty-Three Pages of the Dresden Codex: The Divination Pages." Thesis, Graduate School of the University of Texas, May 1996.

Bartolomé Howel, Juan. "Fontaleza de Santa Teresa." Nomada website. Accessed March 23, 2023.

Bassie-Sweet, Karen. "Corn Deities and the Complementary Male/Female Principle." *La Tercera Mesa Redonda de Palenque,* July 1999. Revised September 2000.

Baudez, Claude-Francois. *Maya Sculpture of Copán: The Iconography.* Norman: University of Oklahoma Press, 2015.

Bauer, Brian S. *Ancient Cuzco: Heartland of the Inca.* Austin, TX: University of Texas Press, 2010.

Begley, Adam. *The Great Nadar: The Man Behind the Camera.* New York: Tim Duggan Books, 2017.

Bishop, Janice L., Daniela Tirsch, Livio L. Tornabene, Ralf Jaumann, Alfred S. McEwen, Patrick C. McGuire, Anouck Ody, et al. "Mineralogy and Morphology of Geologic Units at Libya Montes, Mars: Ancient Aqueously Derived Outcrops, Mafic Flows, Fluvial Features, and Impacts." *Journal of Geophysical Research: Planets* 118, no. 3 (March 31, 2013): 487–513.

Blitz, John H. *Moundville.* Tuscaloosa: University of Alabama Press, 2008.

Bostock, John, and Henry Thomas Riley. *The Natural History of Pliny, Volume 2.* London: H. G. Bohn, 1855. National Library of Medicine website. Accessed March 23, 2023.

Boutwell, Jane. "Anonymous Art." *New Yorker,* November 14, 1964.

Bovill, Carl. *Fractal Geometry in Architecture and Design.* Boston: Birkauser, 1996.

Braakhuis, Edwin. "The Way of All Flesh: Sexual Implications of the Mayan Hunt." *Anthropos* 96, no. 2 (2001): 391–409.

Brandenburg, John E. *Death on Mars: The Discovery of a Planetary Nuclear Massacre.* Kempton, IL: Adventures Unlimited, 2015.

———. "Evidence of a Massive Thermonuclear Explosion on Mars in the Past: The Cydonian Hypothesis and Fermi's Paradox." *AIAA Space Forum,* Long Beach California, September 16, 2016.

Brockhampton Press. *Dictionary of Classical Mythology.* Leicester: Brockhampton Press LTD, 1995.

Brodwin, Erin. "Researchers Won an Award for Figuring Out What Happens in the Brains of People Seeing Jesus in Toast." *Business Insider* website, September 24, 2014. Accessed March 23, 2023.

Burland, Cottie, Irene Nicholson, and Harold Osborne. *Mythology of the Americas.* London: Hamlyn, 1970.

Carmack, Robert M. *The Quiché Mayas of Utatlán: The Evolution of a Highland Guatemala Kingdom.* Norman: University of Oklahoma Press, 1981.

Carwardine, Mark. *Animal Records*. New York: Sterling Publishing Company, Inc., 2008.

Casana, Jesse, and Jackson Cothren. "The CORONA Atlas Project: Orthorectification of CORONA Satellite Imagery and Regional-Scale Archaeological Exploration in the Near East." In *Mapping Archaeological Landscapes from Space. SpringerBriefs in Archaeology*, Vol 5. New York: Springer, 2013.

Cascone, Sarah. "NASA Suggests Aliens May Be Behind Ancient Rock Art" Art World website, May 24, 2014. Accessed March 23, 2023.

Chevalier, Jean, and Alain Gheerbrant. *A Dictionary of Symbols*. London: Penguin Books, 1996.

Coe, Michael D. *Breaking the Maya Code*. New York: Thames and Hudson, 1992.

———. *The Olmec World: Ritual and Rulership*. Princeton: The Art Museum, Princeton University, 1996.

Coe, Michael D., and David Grove. *The Olmec and Their Neighbors: Essays in Memory of Matthew W. Stirling*. Washington, DC: Dumbarton Oaks, 1981.

Coe, Sophie D. and Michael D. Coe. *The True History of Chocolate*. New York: Thames and Hudson, 1996.

Comer, Douglas C., and Michael J. Harrower. *Mapping Archaeological Landscapes from Space*. New York: Springer, 2013.

Craig, Michael J. "Geometrical Features on Mars, in 'Atlantis Chaos.'" Secret Mars, The Mars Archaeology Archive website, August 11, 2019. Accessed March 23, 2023.

Dakss, Brian. "Old Man of the Mountain Collapses." CBS News website, May 3, 2003. Accessed March 23, 2023.

Dale, M. A., George J. Haas, James S. Miller, William R. Saunders, A.J. Cole, Joseph M. Friedlander, and Susan Orosz. "Avian Formation on a South-Facing Slope Along the Northwest Rim of the Argyre Basin." *Journal of Scientific Exploration* 25, no. 3 (September 10, 2011).

Denevan, William M. "The Pristine Myth: The Landscape of the Americas in 1492." *Annals of the Association of American Geographers* 82, no. 3 (September 1992): 369–385.

De Pablo, Miguel Ángel, and Andrea Pacifici. "Geomorphological Evidence of Water Level Changes in Nepenthes Mensae, Mars." *Icarus* 196, no. 2 (August 2008): 667–671.

DeRosa, R. "Parrotopia, ID 97522, Layered Knobs and Rectilinear Ridges in Nereidum Montes." MRO HiRISE website, March 9, 2014. Accessed March 23, 2023.

Diehl, Richard A. *The Olmecs: America's First Civilization*. London: Thames and Hudson, 2004.

Dommermuth-Costa, Carol. *Nikola Tesla: A Spark of Genius*. Minneapolis, MN: Lerner Publishing Group, 1994.

Doniger, Wendy, ed. *Merriam-Webster's Encyclopedia of World Religions*. Springfield, MA: Merriam-Webster, 1999.

Donnell, Clayton. *Breaking the Fortress Line 1914*. Barnsley, South Yorkshire: Pen and Sword, 2013.

Dorrian, Mark, and Frédéric Pousin, eds. *Seeing from Above: The Aerial View in Visual Culture*. London: I. B. Tauris & Co., Ltd., 2013.

European Space Agency. "1st Mars Express Science Conference," ESA Mars Express website, September 1, 2019. Accessed March 23, 2023.

———. "Chaos in Atlantis Basin." ESA website, October 27, 2014. Accessed March 23, 2023.

———. "Launch Phase." ESA Mars Express website, January 29, 2024, Accessed March 23, 2023.

Faust, Wil. "The Anomaly Hunters Roundtable Study: Parrotopia, the Parrot and City Complex." Parrotopia—Anomaly Hunters Roundtable Study website, January 17, 2003. Accessed March 23, 2023.

———. "Parrotopia Update No.1: Was the City a Port?" Parrotopia—Anomaly Hunters Roundtable Study website, (Wayback Machine). January 10, 2003. Accessed March 23, 2023.

Field, Daniel J., Jacques A. Gauthier, Benjamin L. King, Davide Pisani, Tyler R. Lyson, and Kevin J. Peterson. "Toward Consilience in Reptile Phylogeny: miRNAs Support an Archosaur, Not Lepidosaur, Affinity for Turtles." *Evolution & Development* 16, no. 4 (July/August 2014): 189–96.

Fletcher, Valerie J. *Isamu Noguchi Master Sculptor*. London: Scala Publishers, 2005.

Ford, Derek, and Paul Williams. *Karst Hydrogeology and Geomorphology*. Chichester: John Wiley & Sons Ltd., 2007.

Foster, Lynn V. *Handbook to Life in the Ancient Maya World*. Oxford: Oxford University Press, 2005.

Freidel, David, Linda Schele, and Joy Parker. *Maya Cosmos: Three Thousand Years on the Shaman's Path*. New York: Quill, 1995.

Fritz, John M., and George Mitchell. *Hampi Vijayanagara*. Mumbai: Jaico Publishing House, 2015.

Gangal, Kavita, Graeme R. Sarson, and Anvar Shukurov. "The Near-Eastern Roots of the Neolithic in South Asia." *PloS One* 9, no. 5 (2014): e95714.

Gauch Hugh G., and Hugh G. Gauch, Jr. *Scientific Method in Practice*. New York: Cambridge University Press, 2003.

Gayol, Carlos Pallán. "The Many Faces of Chaahk: Exploring the Role of a Complex and Fluid Entity within Myth, Religion and Politics." In "The Maya and Their Sacred Narratives: Text and Context in Maya Mythologies." *Acta Mesoamericana* 20 (2009): 31.

Giles, Sue, and Valerie Harland. "Digitisation of the Adela Breton Collection at Bristol Museum & Art Gallery." ArtUK website, November 13, 2014. Accessed March 23, 2023.

Gipson, Mack and Victor K. Ablordeppey. "Pyramidal Structures on Mars." *Icarus* 22, no. 2 (June 1974): 197–204.

Goldstein, Steven. "Watch What You're Thinking! The Skeptic's Toolbox II Conference." *Skeptical Inquirer* 18, no. 4 (Summer 1994).

Gott, Kendall D. *Where the South Lost the War: An Analysis of the Fort Henry—Fort Donelson Campaign, February 1862.* Mechanicsburg, PA: Stackpole Books, 2003.

Gowin, Emmet, and Robert Adams. *The Nevada Test Site.* Princeton: Princeton University Press, 2019.

Graham, Mark Miller. "Mesoamerican Jade and Costa Rica." In *Jade in Ancient Costa Rica.* New York: The Metropolitan Museum of Art, 1998.

Graves, Robert. *The Greek Myths.* London: Moyer/Bell, 1994.

Greeley, Ronald. *Introduction to Planetary Geomorphology.* Cambridge, UK: Cambridge University Press, 2013.

Greeley, Ronald, Nicholas Lancaster, Steven Lee, and Peter Thomas. "Martian Aeolian Processes, Sediments, and Features." *Mars* (1992).

Greshko, Michael. "See Newly Discovered Ancient Drawings in Peru Desert." *National Geographic* website, April 5, 2018. Accessed March 23, 2023.

Grigoriev, Stanislav A., and Nikolai M. Menshenin. "Discovery of Geoglyphs on the Zjuratkul Ridge in Southern Urals." *Antiquity* 86, no. 331 (March 2012). Accessed March 23, 2023.

Grofe, Michael J. "The Recipe for Rebirth: Cacao as Fish in the Mythology and Symbolism of the Ancient Maya." Department of Native American Studies, University of California at Davis. September 23, 2007.

Grube, Nikolai and Linda Schele. "Kuy, the Owl of Omen and War." *Mexicon* 16, no. 1 (February 1994): 10–17. Accessed March 23, 2023.

Grzimek, B. *Grzimek's Animal Life Encyclopedia*, Volume 9, Birds II. 2nd ed. Detroit: Thomson/Gale, 2003.

Guzman, Irene. "The (Triangular) Office Building Designed by Jo Coenen and Archisquare in Parma, Abitare." Abitare website, June 17, 2015. Accessed March 23, 2023.

Haas, George J. "Keyhole—Exclamation Mark on Mars," The Cydonia Institute Discussion Board, July 17, 2013. Accessed March 23, 2023.

Haas, George J., and William R. Saunders. *The Cydonia Codex: Reflections from Mars.* Berkeley, CA: Frog Ltd., 2005.

———. *The Martian Codex: More Reflections from Mars.* Berkeley, CA: North Atlantic Books, 2009.

Hartley's Safaris, "Sculpture in Chyulu Hills Kenya Seen from Space." Hartley's Safaris "Latest News" website, March 11, 2012. Accessed March 23, 2023.

Haydon, F. Stansbury. *Military Ballooning During the Early Civil War.* Baltimore: Johns Hopkins University Press, 1941.

Hellmuth, Nicholas Matthew. "Structure 5D-73, Burial 196, Tikal, Peten, Guatemala: A Preliminary Report." BA Honors Thesis. Foundation for Latin American Anthropological Research. March 31, 1967.

Heritage Gateway. "Longstones Barrow." Historic England Research Records website. Accessed March 23, 2023.

Herrle, Peter and Erik Wegerhoff, eds. "Longstones Barrow" In *Architecture and Identity*, Volume 9, Habitat—International LIT, Verlag Münster, 2008.

Hiesinger Harold, and James W. Head, III. "Topography and Morphology of the Argyre Basin, Mars: Implications for Its Geologic and Hydrologic History." *Planetary and Space Science*, 50, no. 10–11 (August–September 2002): 939–981.

Higyou. "Several Small Modern Buildings following an Arrow-Shaped One, 3D, Isolated." Image ID: MH45F5. Alamy website, May 1, 2018. Accessed March 23, 2023.

Holm, Bill. *Northwest Coast Indian Art: an Analysis of Form.* Seattle: University of Washington Press, 1965.

Homer. *The Odyssey with an English Translation by A. T. Murray, Ph.D., Volume 1 & 2.* Trans. A. T. Murray. Cambridge, MA: Harvard University Press; London: William Heinemann, Ltd., 1919.

Howell, Elizabeth. "InSight Lander: Probing the Martian Interior," Space.com website, November 26, 2018. Accessed March 23, 2023.

———. "Mariner 9: First Spacecraft to Orbit Mars." Space.com website, November 8, 2018. Accessed March 23, 2023.

Hrala, Josh. "Stephen Hawking Warns Us to Stop Reaching Out to Aliens before It's Too Late." Science Alert website, November 4, 2016.

Ishihara, Reiko, Karl A. Taube, and Jaime J. Awe. "The Water Lily Serpent Stucco Masks at Caracol, Archaeological Investigations in the Eastern Maya Lowlands: Papers of the 2005 Belize Archaeology Symposium, Belize." *Research Reports in Belizean Archaeology* 3 (2006).

Jankowski, Paul. *Verdun: The Longest Battle of the Great War.* Oxford: Oxford University Press, 2014.

Jarus, Owen. "Mysterious Symbols in Kazakhstan: How Old Are They, Really?" Live Science website, November 5, 2015.

———. "Nazca Lines of Kazakhstan: More Than 50 Geoglyphs Discovered." Soul:Ask website, September 26, 2014. Accessed March 23, 2023.

———. "Visible Only from Above, Mystifying 'Nazca Lines' Discovered in Mideast." NBC Science News website, September 15, 2011.

Jet Propulsion Laboratory. "A Regional View of the Libya Montes," NASA's Mars Images website, May 22, 2000. Accessed March 23, 2023.

Jobson, Christopher. "WISH: A Monumental 11-Acre Portrait in Belfast by Jorge Rodríguez-Gerada." Colossal website, October 20, 2013. Accessed March 23, 2023.

Kawagoe, Aileen. "Towering Tumuli of the Kofun Era." *Heritage of Japan* (blog). 2013.

Keener, Candis M. "The Baby Jaguar Series: A Comparative Analysis." MA thesis, Kent State University, 2009.

Kelly, John. "Hard-Core 'Star Wars' Fans Dress Up and Troop Around." *Washington Post* website, December 1, 2013. Accessed March 23, 2023.

Kelly, Joyce. *An Archaeological Guide to North Central America: Belize, Guatemala, Honduras, and El Salvador*. Norman: University of Oklahoma Press, 1996.

Kennedy, David L. "Desktop Archeology." *Aramco World* 60, no. 4 (July/August 2009).

Kessler, Rebecca. "Thousands of Tombs in Saudi Desert Spotted from Space." Live Science website, February 15, 2011.

Kidder, Tristram R. "Excavations at the Jordan Site (16MO1), Morehouse Parish, Louisiana." *Southeastern Archaeology* 11, no. 2 (Winter 1992): 109–131.

Kiefer, Walter S., Allan H. Treiman, and Stephen M. Clifford. "The Red Planet: A Survey of Mars," 2nd ed. Slide 15. Lunar and Planetary Institute website, 2011. Accessed March 23, 2023.

Kleber, John E. *The Kentucky Encyclopedia*. Lexington: University Press of Kentucky, 2014.

Kuilman, Marten. "Quadralectic Architecture—A Panoramic Review." *Quadralectic Architecture* (blog), August 26, 2013. Accessed March 23, 2023.

Lambert, W. G. "The God Aššur." *Iraq* 45, no. 1 (Spring 1983): 82–86. Accessed March 21, 2023.

Lange, Fredrick W. *Pre-Columbian Jade: New Geological and Cultural Interpolations*. Salt Lake City: University of Utah Press, 1993.

Lawrence Livermore National Laboratory, Weapons and Complex Integration, "Big Facility, Explosives Experimental Facility." LLNL website. Accessed March 23, 2023.

Laylin, Tafline. "Agora Tower: Twisting Skyscraper Wrapped with Vertical Gardens Breaks Ground in Taipei." Inhabitat website, March 10, 2013. Accessed March 23, 2023.

———. "Luis de Garrido Unveils Incredible Glass-Domed Eco House Shaped Like the Eye of Horus." Inhabitat website, September 17, 2011. Accessed March 23, 2023.

Leaman, D. E. "The Nature of Jointing, Tasman Peninsula, Southern Tasmania." *Papers and Proceedings of the Royal Society of Tasmania* 133, no. 1 (1999): 65–76.

Leung, Clint. *An Overview of Pacific Northwest Native Indian Art, Exquisite Pacific Northwest Native Indian & Inuit Art.* Free Sprit Gallery, 2006.

Levasseur, John P. "Hypothesis," Artifacts on Mars website, 2000. Accessed March 23, 2023.

Levasseur, John P., George J. Haas, William R. Saunders, and Horace Crater. "Analysis of the MGS and MRO Images of the Syria Planum Profile Face on Planet Mars." *Journal of Space Exploration* 3, no. 3 (December 30, 2014): 213–230.

Levi, Wendell M. *The Pigeon*, Columbia, SC: R. L. Bryan Company, 1941.

Levine, Yasha. *Surveillance Valley, The Secret Military History of the Internet.* New York: PublicAffairs, 2018.

Levine, Yasha. "Oakland emails give another glimpse into the Google-Military-Surveillance Complex." PandoDaily website, March 7, 2014.

Liebs, Chester H. "Inventory—Nomination form Old Red Mill." National Register Historic Places website, May 8, 1972. Accessed March 23, 2023.

Longhena, Maria, and Walter Alva. *The Incas and Other Andean Civilizations.* San Diego: Thunder Bay Press, 1999.

Longman, Green, Longman & Roberts. "Association of Medical Officers of Asylums and Hospitals for the Insane (London), Medico-Psychological Association of Great Britain and Ireland, Royal Medico-Psychological Association, Harvard University." *The Journal of Mental Science* 13 (1868).

Malin Space Science Systems, "West Argyre, MGS MOC Release No. M0C2-1191." Mars Global Surveyor, Mars Orbiter Camera website, August 22, 2005. Accessed March 23, 2023.

Mamiya, Yasuyuki, Yoshiyuki Nishio, Hiroyuki Watanabe, Kayoko Yokoi, Makoto Uchiyama, Toru Baba, and Osamu Iizuka, et al. "The Pareidolia Test: A Simple Neuropsychological Test Measuring Visual Hallucination-Like Illusions." *Plos One* 11, no. 5 (2016): e0154713. Accessed March 23, 2023.

Manney, Kevin. "Tiny Tech Company Awes Viewers." *USA Today*, March 21, 2003.

Mao, Frances. "Marree Man: The Enduring Mystery of a Giant Outback Figure." *BBC News* website, June 26, 2018.

Marshall, Steve. "Exploring Avebury: The Essential Guide." *The Antiquaries Journal* 97 (September 2017): 314–315.

Mars Viewer, ESA, Mars Express, H2081_0000_ND3. Mars Space Flight Facility, Arizona State University Mars Express website, August 28, 2005. Accessed March 23, 2023.

Mars Viewer, ESA, Mars Express, H2004_0000_ND3. Arizona State University Mars Space Flight Facility website (Mars Express HRSC/SRC), August 6, 2005. Accessed March 23, 2023.

Mars Viewer, ESA, Mars Express, H5208_0000_ND3. Arizona State University Mars Space Flight Facility website (Mars Express HRSC/SRC), January 21, 2008. Accessed March 23, 2023.

Mars Viewer, Mars Odyssey THEMIS, V18218024. Arizona State University Mars Space Flight Facility website (THEMIS), January 22, 2006. Accessed March 23, 2023.

Mars Viewer, Mars Odyssey, THEMIS, V26406033. Arizona State University Mars Space Flight Facility website (THEMIS), November 27, 2007. Accessed March 23, 2023.

Mars Viewer, Mars Odyssey, THEMIS, V13700005. Arizona State University Mars Space Flight Facility website (THEMIS), January 15, 2005. Accessed March 23, 2023.

Mars Viewer, Mars Odyssey, THEMIS, V61919005. Arizona State University Mars Space Flight Facility website (THEMIS), November 29, 2015. Accessed March 23, 2023.

Mars Viewer, MOC M1402185. "S-Facing Slope on Massif in W Argyre Rim Region." Arizona State University Mars Space Flight Facility website (Mars Orbiter Camera), April 30, 2000. Accessed March 23, 2023.

Mars Viewer, MOC M0903566. "Contact between Acheron Fossae Upland and NE Amazonis Plains." Arizona State University Mars Space Flight Facility website (Mars Orbiter Camera), November 14, 1999. Accessed March 23, 2023.

Mars Viewer, MOC S13-01480. "Repeat Layered Material and Rectilinear Ridges in M14-02185." Arizona State University Mars Space Flight Facility website (Mars Orbiter Camera), December 15, 2005. Accessed March 23, 2023.

Mars Viewer, MRO HiRISE CTX B03_010626_1821_XN_02N228W. Arizona State University Mars Space Flight Facility website (MRO Context Camera), November 1, 2008. Accessed March 23, 2023.

Mars Viewer, MRO HiRISE CTX B17_016270_1878_XN_07N236W. "Nepenthes Mensae." Arizona State University Mars Space Flight Facility website (MRO Context Camera), January 1, 2010. Accessed March 23, 2023.

Mars Viewer, MRO HiRISE CTX B20_017574_1965_XN_16N198W. "Ride-Along with HiRISE." Arizona State University Mars Space Flight Facility website (MRO Context Camera), April 26, 2010. Accessed March 23, 2023.

Mars Viewer, MRO HiRISE CTX D06_029600_1968_XN_16N198W. "Ride-Along with HiRISE." Arizona State University Mars Space Flight Facility website (MRO Context Camera), November 18, 2012. Accessed March 23, 2023.

Mars Viewer, MRO HiRISE CTX D15_033142_1288_XI_51S054W. "Western Argyre." Arizona State University Mars Space Flight Facility website (MRO Context Camera), August 21, 2013. Accessed March 23, 2023.

Mars Viewer, MRO HiRISE CTX F16_041969_1960_XN_16N198W. "Ride-Along with HiRISE." Arizona State University Mars Space Flight Facility website (MRO Context Camera), July 10, 2015. Accessed March 23, 2023.

Mars Viewer, MRO HiRISE CTX G01_018708_1959_XN_15N198W13. "Ride-Along with HiRISE." Arizona State University Mars Space Flight Facility website (MRO Context Camera), July 24, 2010. Accessed March 23, 2023.

Mars Viewer, MRO HiRISE CTX G21_026542_1894_XN_09N235W. "Nepenthes Planum." Arizona State University Mars Space Flight Facility website (MRO Context Camera). March 25, 2012. Accessed March 23, 2023.

Mars Viewer, MRO HiRISE CTX J03_046136_1965_XN_16N198W. "Ride-Along with HiRISE." Arizona State University Mars Space Flight Facility website (MRO Context Camera). May 30, 2016. Accessed March 23, 2023.

Mars viewer, MRO HiRISE CTX J04_046177_1871_XN_07N235W. "Landforms in the Nepenthes Mensae Region." Arizona State University Mars Space Flight Facility website (MRO Context Camera), June 2, 2016. Accessed March 23, 2023.

Mars Viewer, MRO HiRISE CTX K05_055473_1871_XN_07N236W. "Landforms in the Nepenthes Mensae Region." Arizona State University Mars Space Flight Facility website (MRO Context Camera), May 27, 2018. Accessed March 23, 2023.

Mars Viewer, MRO HiRISE CTX K11_057768_1854_XN_05N235W. "Nepenthes Mensae." Arizona State University Mars Space Flight Facility website (MRO Context Camera), November 22, 2018. Accessed March 23, 2023.

Mars viewer. MRO HiRISE CTX N01_062858_1871_XN_07N236W. "Landforms in the Nepenthes Mensae Region." Arizona State University Mars Space Flight Facility website (MRO Context Camera), December 24, 2019. Accessed March 23, 2023.

Mars Viewer, MRO HiRISE CTX N02_063214_1882_XN_08N236W. "Landforms in the Nepenthes Mensae Region." Arizona State University Mars Space Flight Facility website (MRO Context Camera), January 21, 2020. Accessed March 23, 2023.

Mars Viewer, MRO HiRISE CTX P03_002318_1961_XN_16N198W. "Scoured Terrain and Associated Landforms in East/Northeast Cerberus Region." Arizona State University Mars Space Flight Facility website (MRO Context Camera), January 24, 2007. Accessed March 23, 2023.

Mars Viewer, MRO HiRISE CTX P11_005219_1961_XN_16N198W. "Landforms in Cerberus Region." Arizona State University Mars Space Flight Facility website (MRO Context Camera), September 7, 2007. Accessed March 23, 2023.

Mars Viewer, MRO HiRISE CTX P13_006142_1964_XN_16N198W. "Landforms in Cerberus Region." Arizona State University Mars Space Flight Facility website (MRO Context Camera), November 18, 2007. Accessed March 23, 2023.

Mars Viewer, MRO HiRISE CTX P14_006672_1836_XN_03N267W. "Libya Montes." Arizona State University Mars Space Flight Facility website (MRO Context Camera), December 29, 2007. Accessed March 23, 2023.

Mars Viewer, Viking 883A03. "Viking Orbiter 1." Arizona State University Mars Space Flight Facility website (Viking), November 17, 1978. Accessed March 23, 2023.

Martin, Simon, and Nikolai Grube. *Chronicle of the Maya Kings and Queens: Deciphering the Dynasties of the Ancient Maya*. London: Thames and Hudson, 2000.

Mayer, Emile. "Brialmont." *Revue d'histoire modern* 2, no. 9, 1927. Accessed March 23, 2023.

McDonald, David. *The Encyclopedia of Mammals*. Oxford University Press, 2010.

McEwen, Alfred S. "Exclamation Mark on Mars." University of Arizona LPL HiRISE website, January 12, 2011. Accessed March 23, 2023.

———. "Faculty." The University of Arizona, Lunar and Planetary Laboratory website.

———. "Instruments." NASA website (Mars Reconnaissance Orbiter). Accessed March 23, 2023.

———. MRO HiRISE ESP_020794_1860. "Exclamation Mark on Mars." University of Arizona HiRISE website. January 2, 2011. Accessed March 23, 2023.

Mckeever, R., W. B. Whalley, B. J. Smith. "Basalt Weathering in a Temperate Marine Environment." *Developments in Soil Science* 19 (1990): 519–523.

McLachlan, W. G. "The Mounds of Lake Waubesa Region." *The Wisconsin Archeologist* 11–13, Wisconsin Natural History Society. *University of Chicago Wisconsin Archeological Society*, 12, no. 4 (January 1914).

Mean Green. "Another Interesting Area WITH Possible Structures/Ruins ;-)" Facebook, April 26, 2018.

Megahed, Naglaa A. "Origami Folding and Its Potential for Architecture Students." *The Design Journal: An International Journal for All Aspects of Design* 20, no. 2 (January 16, 2017): 279–297.

Mieczkowski, Yanek. *Eisenhower's Sputnik Moment: The Race for Space and World Prestige*. Ithaca: Cornell University Press, 2013.

Milbrath, Susan. *Star Gods of the Maya Astronomy in Art, Folklore and Calendars*, Austin: University of Texas Press, 1999.

Miller, Arthur G. *The Codex Nuttall*. Ed. Zelia Nuttall. New York: Dover, 1975.

Miller, Jim. "The Anomaly Hunters Roundtable Study: Parrotopia, the Parrot and City Complex." Anomaly Hunters website, January 18, 2003.

Miller, Mary. "The Willfulness of Art: The Case of Bonampak." *RES: Anthropology and Aesthetics* 42 (Autumn 2002).

Miller, Mary Ellen. "A Re-Examination of the Mesoamerican Chacmool." *The Art Bulletin* 67, no. 1 (March 1985): 7–17.

Miller, Mary, and Karl Taube. *An Illustrated Dictionary of The Gods and Symbols of Ancient Mexico and the Maya.* New York: Thames and Hudson, 2015.

Mitchell, Spencer T. "Visual Communications of Power: The Iconography of the Classic Maya Naranjo-Sa'al Polity." MA thesis, Texas Tech University, December 2016.

Montgomery, John. *Dictionary of Maya Hieroglyphics.* New York: Hippocrene Books, Inc., 2002.

Mora, Marlin Calvo. *Gold, Jade, Forests: Costa Rica.* University of Washington Press, 1995.

Morishima, K., Mitsuaki Kuno, Akira Nishio, Nobuko Kitagawa, Yuta Manabe, Masaki Moto, and Fumihiko Takasaki. "Discovery of a Big Void in Khufu's Pyramid by Observation of Cosmic-Ray Muons." *Nature* 552 (2017): 386–390.

Morley, Sylvanus G. "The Inscriptions of Naranjo, Northern Guatemala." *American Anthropologist* 11, no. 4 (October–December 1909): 543–562.

Morton, Ella. "The Strange Story of Australia's Mysterious Marree Man." Atlas Obscura on Slate website, January 26, 2015.

Muniz, Vik. "Pictures of Earthwork." Nichido Contemporary Art website, 2009. National Park Service, "How Big Are the Heads?" National Park Service Website for Mount Rushmore National Memorial, South Dakota, April 2, 2020. Accessed March 23, 2023.

NASA. "Mars Global Surveyor." Mars Global Surveyor Home Page, NASA website. Accessed November 23, 2023.

NASA. "Nepenthes Mensae." Jet Propulsion Laboratory Image website, April 11, 2012. Accessed March 23, 2023.

Nava, Pedro Sueldo. *A Walking Tour of Machupicchu: Editorial de Cultura Andina.* Austin: University of Texas, 1976.

News.com.au. "Dick Smith is Offering a $5000 Reward for Anyone Who Can Trace the Origins of the Marree Man." News.com.au website, June 26, 2018.

Newsgram Desk. "UNESCO Declares Al 'Ula as Saudi's First World Heritage Site: The 2,000-Year-Old Town Is Made of Mud and Stone." June 14, 2016. Accessed March 23, 2023.

Newton, Alfred. *A Dictionary of Birds.* London: A. and C. Black, 1893.

Nussbaumer, J. "Possible Sea Sediments due to Glaciofluvial Activity in Elysium Planitia, Mars." Paper presented at European Planetary Science Congress, Berlin, Germany, September 2006.

O'Callaghan, Jonathan. "Signs of Recent Volcanic Eruption on Mars Hint at Habitats for Life." *New York Times* website (Science), November 20, 2020. Accessed March 23, 2023.

Ohno, Susumu. *Sex Chromosomes and Sex-Linked Genes, Volume 1, Monographs on Endocrinology*. Berlin: Springer Science & Business Media, 2013.

Orejel, Jorge L. "The 'Axe/Comb' Glyph as Ch'ak." *Reports on Ancient Maya Writing*, 31. Washington, DC: Center for Maya Research, 1990.

Oriental Institute. "Persepolis and Ancient Iran, Aerial Survey Flights, 1935." University of Chicago Institute for the Study of Ancient Cultures website. Accessed March 23, 2023.

Orme, Greg. "The Kings Valley, Mars, Why We Must Go to Mars," Facebook, July 17, 2019. Accessed March 23, 2023.

Ouellette, Jennifer, and Ars Technica. "Why Humans See Faces in Everyday Objects: The Ability to Spot Jesus' Mug in a Piece of Burnt Toast Might Be a Product of Evolution." *Wired* website, July 14, 2021. Accessed March 23, 2023.

Parcak, Sarah. *Satellite Remote Sensing for Archaeology*. New York: Routledge, 2009.

Parks, R. J. "Mariner 9 and the Exploration of Mars." *Astronautical Research 1972* (1973): 149–162.

Payne, Roz. "Dove in Pentagon." Sixties Archive website.

Peet, Stephen Denison. *State Historical Society of Wisconsin, Collections—State Historical Society of Wisconsin*. The University of California, General Books LLC, 1882.

Perez, Chris. "Scientists Discover Hidden Chamber in Great Pyramid." *New York Post* website, November 2, 2017. Accessed March 23, 2023.

Phillips, Adam. "Interview with Ground Zero Architect Daniel Libeskind—2003-03-23," *VOA News* website, October 26, 2009. Accessed March 23, 2023.

Phillips, Charles. *The Complete Illustrated History of the Aztec & Maya*. China: Hermes House, 2010.

Phillips, Erica E. "Earthlings Look for Signs in New Photos of Mars." *Wall Street Journal* CCLX, no. 43 (August 21, 2012).

Piccaluga, Ennio. "The Cities of Mars." *Epiccaluga Wordpress* (blog). October 12, 2019. Accessed March 23, 2023.

Pitts, Mark. *Maya Numbers and the Maya Calendar: A Non-Technical Introduction to Maya Glyphs—Maya Book 2*. Foundation for the Advancement of Mesoamerican Studies (FAMSI) website, 2009.

Powell, Sian. "Marree Man Refuses to Divulge His Secret." *The Australian* website, June 25, 2018.

Pozos, Randolfo Rafael. *The Face on Mars: Evidence for a lost Civilization?* Chicago: Chicago Review Press, 1986.

Pringle, Heather. "Satellite Imagery Uncovers Up to 17 Lost Egyptian Pyramids." *Science* website, May 27, 2011. Accessed March 23, 2023.

Protzen, Jean-Pierre, and Stella Nair. "Who Taught the Inca Stonemasons Their Skills? A Comparison of Tiahuanaco and Inca Cut-Stone Masonry." *Journal of the Society of Architectural Historians* 56, no. 2 (June 1997): 146–167.

Raza, Javed (A.K.A. Jay Raza). "Archaeology on Mars Series . . . Discovery of an Earth Like 'City' & Structures," Facebook, July 27, 2019. Accessed March 23, 2023.

Reader's Digest Association. *The World's Last Mysteries.* Pleasantville, NY: Reader's Digest Association, 1978.

———. "A 'City' on MARS in the Atlantis Chaos Region." Facebook, July 24, 2019. Accessed March 23, 2023.

Recinos, Adrian. *Popol Vuh: The Sacred Book of the Ancient Quiché Maya.* Civilization of the American Indian Series. Translated by Delia Goetz and Sylvanus Griswold Morley. Norman: University of Oklahoma Press, 1950.

Reed, Alma M. *The Ancient Past of Mexico.* New York: Crown Publishers Inc., 1966.

Rogers, Andrew. "Lion's Paw." Andrewrogers.org website (Land Art/Kenya), 2010. Accessed March 23, 2023.

Ross, John F. "First City in the New World? Peru's Caral Suggests Civilization Emerged in the Americas 1,000 Years Earlier Than Experts Believed." *Smithsonian Magazine* 33, no. 6 (August 2002).

Rubin, Josh. "2016 TED Prize Winner: Dr Sarah Parcak's Crowdsourcing Space Archaeology." Cool Hunting website, February 17, 2016. Accessed March 23, 2023.

Sacks, David. *A Dictionary of the Ancient Greek World.* New York: Oxford University Press US, 1997.

Sagan, Carl. "Christmas Lectures: The Planets: Mars before Viking." Royal Institution, London 1977. Accessed March 23, 2023.

———. *Cosmos.* New York: Random House, 1980.

———. *Cosmos.* Episode 12, "Encyclopaedia Galactica." Aired December 14, 1980.

Sagan, Carl, and Ann Druyan. *The Demon-Haunted World: Science as a Candle in the Dark.* New York: Random House Publishing Group, 1997.

Saunders, Nicholas J. "Predators of Culture: Jaguar Symbolism and Mesoamerican Elites." *World Archaeology* 26, no. 1 (June 1994): 104–117.

Saunders, William R. "Mean City: Darkness on the Edge of Forever." The Cydonia Institute Discussion Board, May 18, 2018. Accessed March 23, 2023.

Saunders, William R., George J. Haas, and James Miller. "A Persistent Avian Formation on a South-Facing Slope, along the Northwest Rim of the Argyre Basin of Mars." *Journal of Scientific Exploration* 36, no. 2 (Summer 2022).

Saunders, William R., George J. Haas, James Miller, and Michael Dale. "A Wedge and Dome Formation Set within the Flat Plains of Libya Montes." *Journal of Space Exploration* 4, no. 3 (November 17, 2016).

Schele, Linda. "The Universe: Now and Beyond." NASA Administrator's Third Seminar Series, NASA News website, March 13, 1995. Accessed March 23, 2023.

Schele, Linda, and David Freidel. *A Forest of Kings: The Untold Story of the Ancient Maya*. New York: Quill, 1990.

Schele, Linda, and Peter Mathews. *The Code of Kings: The Language of Seven Sacred Maya Temples and Tombs*. New York: Touchstone Books/Simon & Schuster, 1999.

Scherz, James P., and Buck Trawicky. "Survey Report Hudson Park Mound Group—Dane County Madison Wisconsin." Ancient America website, March 24, 2017. Accessed March 23, 2023.

Schirber, Michael. "Attempts to Contact Aliens Date Back More Than 150 Years." Space.com website, January 29, 2009.

Schnitzler, Hans-Ulrich, Elisabeth K. V. Kalko, Ingrid Kaipf, and Alan D. Grinnell. "Fishing and Echolocation Behavior of the Greater Bulldog Bat, *Noctilio leporinus*, in the Field." *Behavioral Ecology and Sociobiology* 35, no. 5 (November 1994).

Schoch, Robert. *Ancient Aliens*. Season 5, episode 6, "Secrets of the Tombs." Aired January 25, 2013, on History Channel.

Scientific American. "The Formation of Mountains by Water, the Influence of Erosion by Water in Modeling the Landscape." *Scientific American Supplement* LXXIL, No. 1859, New York, August 19, 1911.

Seidler, Harry. "Riverside Center," Harry Seidler & Associates website (Offices and Public Buildings), 1986. Accessed March 23, 2023.

Shayne, Tasha. "The Enigma of the Lost Chinese Pyramids of Xi'an." Gaia website, November 12, 2019. Accessed March 23, 2023.

Siddiqi, Asif A. "Deep Space Chronicle: A Chronology of Deep Space and Planetary Probes, 1958–2000." *Monographs in Aerospace History*, no. 24 (June 2002).

Simmons, Nancy B. *Mammal Species of the World: A Taxonomic and Geographic Reference*. 3rd ed. Baltimore: Johns Hopkins University Press, 2005.

Simpsons, The. Season 1, episode 8, "The Telltale Head." Aired February 25, 1990, on Fox. Accessed March 23, 2023.

Sitchin, Zecharia. *The 12th Planet: Book I of The Earth Chronicles*. New York: Avon Books, 1976.

———. *Genesis Revisited: Is Modern Science Catching up with Ancient Knowledge?* New York: Avon Books, 1990.

———. *The Lost Book of Enki: Memoirs and Prophecies of an Extraterrestrial God*. Rochester, VT: Bear & Company, 2002.

———. *The Lost Realms: Book IV of The Earth Chronicles*. New York: Avon Books, 1990.

———. *The Wars of Gods and Men: Book III of The Earth Chronicles*. New York: Avon Books, 1985.

Skele, Mike. "The Great Knob: Interpretations of Monks Mound." *Studies in Illinois Archaeology*, no. 4. Springfield: Illinois Historic Preservation Agency, 1988.

Smith, I. F. and J. G. Evans, "Excavation of Two Long Barrows in North Wiltshire." *Antiquity* 42, no. 166 (June 1, 1968): 138–142.

Solis, Felipe, and Ted Leyenaar. *Mexico: Journey to the Land of the Gods*. Amsterdam, Netherlands: Lund Humphries, 2002.

Solis, Ruth S., Jonathan Haas, and Winifred Creamer. "Dating Caral a Pre-Ceramic Site in Supe Valley on the Central Coast of Peru." *Science* 292, no. 5517 (May 2001): 723–6.

Sparavigna, Amelia Carolina. "A Prehistoric Solar Observatory in the Middle of Sahara Desert." *PHILICA*, Article 428 (October 22, 2014).

Spiteri, Stephen C. "Illustrated Glossary of Terms Used in Military Architecture Terms." *Arx: International Journal of Military Architecture and Fortification* (2010): 635–650.

Squier, Ephraim G. and Edwin H. Davis. *Contributions to Knowledge Vol 1.*, Washington, DC: Smithsonian Institution, 1848.

St. Fleur, Nicholas. "Desktop Archaeology: In the Saudi Desert, 400 Gates to the Past." *New York Times*, Oct 24, 2017.

Stevens, Michelle N. "Archaeological Preservation and Environmental Conservation in Arizona's Cienega Valle." *Archaeology Southwest* 15, no. 4 (Fall 2011).

Stuart, David. *The Inscriptions from Temple XIX at Palenque*. San Francisco: Pre-Columbian Art Research Institute, 2005.

Stukeley, William. *Abury, a Temple of the British Druids, with Some Others, Described. Volume 2*. London: Bookfellers, 1743.

———. "Avebury a Present from the Past: The Longstones." Avebury UK website. Accessed March 23, 2023.

Tampke, Jürgen. *The Germans in Australia*. Port Melbourne, AUS: Cambridge University Press, Port Melbourne, 2006.

Tarter, Jill C. *Alien Encounters*. Season 1, episode 1, "Part 1: The Message." Aired March 13, 2012, on Discovery Channel. Accessed March 23, 2023.

Taube, Karl Andreas. *The Major Gods of Ancient Yucatan*. Washington, DC: Dumbarton Oaks Research Library, 1992.

Tedlock, Dennis. *Popol Vuh: the Definitive Edition of the Mayan Book of the Dawn of Life and the Glories of Gods and Kings*. New York: Touchstone, 1986.

Thomas, David Hurst. *Exploring Ancient Native America: An Archaeological Guide*. New York: Macmillan, 1994.

Thompson, J. Eric S. *Maya Hieroglyphic Writing.* Norman: University of Oklahoma Press, 1971.

Thompson, Neville. "Javed Raza's-Atlantis Chassis-ESP 019103 1460." *Gigapan* website, August 7, 2019. Accessed March 23, 2023.

Thompson, Robert Jr. "Linda Schele, Pioneer in the Study of Mayans, Dies at 55." *New York Times*, April 22, 1998.

Toman, William J. "Lizard Effigy Mound, 500–1000 A.D." The Historical Marker Database website, July 16, 2010. Accessed March 23, 2023.

Townsend, Richard F. *Hero, Hawk, and Open Hand: American Indian Art of the Ancient Midwest and South.* Chicago: Art Institute of Chicago in association with Yale University Press, 2004.

Tozzer, Alfred Marston and Glover Morrill Allen. "Animal Figs in the Maya Codices." Volume IV, no. 3 of *Papers of the Peabody Museum of American Archaeology and Ethnology.* Cambridge, MA: Harvard University/Harvard Museum, 1910.

Tuan, Minh. "European Investors Eye US$1 Billion Logistics Project in Vietnam." *The Saigon Times*, September 20, 2020. Accessed March 23, 2023.

Turnbull, Alex. "The Great White Pyramid of China." Google Sight Seeing website, August 20, 2007. Accessed March 23, 2023.

United Press. "U.S. Flier Reports Huge Chinese Pyramid in Isolated Mountains Southwest of Sian." *New York Times*, March 3, 1947.

University of Arizona. MRO HiRISE, ESP_019103_1460 "Iron and Magnesium Clays in and near Atlantis Chaos." University of Arizona HiRISE website, August 24, 2010. Accessed March 23, 2023.

———. MRO HiRISE, ESP_067824_1320. "Layered Knobs and Rectilinear Ridges in Nereidum Montes." University of Arizona HiRISE website, January 14, 2021. Accessed March 23, 2023.

Urcid, Javier. *Zapotec Writing: Knowledge, Power, and Memory in Ancient Oaxaca.* Department of Anthropology, Brandeis University, May 2005. Accessed March 23, 2023.

USGS. "Argyre Planitia." Astrogeology Science Center, Gazetteer of Planetary Nomenclature, Mars, 2022. Accessed March 23, 2023.

———. "Atlantis Chaos." Astrogeology Science Center, Gazetteer of Planetary Nomenclature, Mars, 2022. Accessed March 23, 2023.

———. "Elysium Planitia." Astrogeology Science Center, Gazetteer of Planetary Nomenclature, Mars, 2022. Accessed March 23, 2023.

———. "Libya Montes." Astrogeology Science Center, Gazetteer of Planetary Nomenclature, Mars, 2022. Accessed March 23, 2023.

———. "Nepenthes Mensae." Astrogeology Science Center, Gazetteer of Planetary Nomenclature, Mars, 2020. Accessed March 23, 2023.

Vakoch, Douglas A. *Archaeology, Anthropology, and Interstellar Communication.* The NASA History Series. Createspace Independent Publishing Platform, 2014.

Valenciano, A., M. A. de Pablo, A. Pacifici. "The Role of Water on the Evolution of the Nepenthes Mensa Region of Mars." 40th Lunar and Planetary Science Conference, March 2009.

Van Flandern, Tom. "Preliminary Analysis of April 5 Cydonia Image from the Mars Global Surveyor Spacecraft." Meta Research website, April 10, 1998.

Vick, Charles P. "LACOSSE/ONYX: Radar Imaging Reconnaissance Satellite." Global Security website, July 2005. Accessed March 23, 2023.

Wall Street Journal. "Google Acquires Keyhole: Digital-Mapping Software Used by CNN in Iraq War." *Wall Street Journal*, October 27, 2004. Accessed March 23, 2023.

Waybourn, Marilu. *Images of America: Aztec.* Charleston, SC: Arcadia Publishing 2011.

Weed, William. S. "Tower of Power: What Happened to Engineer Greg Leyh When He Climbed to the Top of a Giant Electrical Transformer That Throws Lightning Bolts into the Sky?—Physical—Electrum, World's Largest Tesla Coil—Cover Story." Lightning on Demand website (Current Science), January 2, 2004.

What on Earth? Season 3, episode 2, "Mystery in the Outback." Aired November 22, 2016, on Science Channel.

White, Max E. *The Archaeology and History of the Native Georgia Tribes.* Gainesville: University Press of Florida, 2002.

Wicks, Mike. "What Does Accent Inns and the Royal Canadian Navy Have in Common?" *Accent Inns* (blog), 2017. Accessed March 23, 2023.

Wilcox, Kevin. "Ambitious mission required extensive technical innovation." *This Month in NASA History* (blog), August 8, 2022. Accessed March 23, 2023.

Wilford, John Noble. "Mapping Ancient Civilization, in a Matter of Days." *New York Times*, May 10, 2010. Accessed March 23, 2023.

Wilford, John Noble. "On the Trail from the Sky: Roads Point to a Lost City," *New York Times*, February 5, 1992.

Williams, Dr. David R. "Viking Mission to Mars," NASA website (Planetary/Viking), April 12, 2018. Accessed March 23, 2023.

Wyatt, Daisy. "Eleven Acre Land Art Unveiled in Belfast." *Independent* website, October 18, 2013. Accessed March 23, 2023.

Yaeger, Jason. "Untangling the Ties That Bind: The City, the Countryside, and the Nature of Maya Urbanism at Xunantunich, Belize." In *The Social Construction of Ancient Cities.* Edited by Monica L. Smith. Washington, DC: Smithsonian Institution Press, 2003.

Yare, Brian. "The Middle Kingdom Egyptian Fortresses in Nubia." Yare website, January 28, 2001. Accessed March 23, 2023.

Young, Biloine W., and Melvin Leo Fowler. *Cahokia: The Great Native American Metropolis*. Urbana: University of Illinois Press, 2000.

Zender, Marc. "Teasing the Turtle from Its Shell: AHK and MAHK in Maya Writing." *PARI Journal* VI, no. 3 (Winter 2006): 1–4.

Index

Note: Page numbers in *italics* refer to illustrations.

About the Author

GEORGE J. HAAS is the founder and premier investigator of the Mars research group known as The Cydonia Institute and is a member of the Society for Planetary SETI Research (SPSR). His research encompasses more than thirty years of study and analysis of NASA and ESA photographs of Mars.

His early schooling was in the visual arts. He was an art instructor, writer, curator, and the former director of the Sculptors' Association of New Jersey. During the 1980s he exhibited extensively throughout the New Jersey and New York area; he was represented by the Grace Harkin Gallery in New York's East Village and had a one-man show in 1989 at the OK Harris Gallery of Art in Soho.

Over the last three decades Haas has studied the art and iconography of North and South American cultures such as the Olmec, Maya, and Aztec. He has been a member of the Pre-Columbian Societies at both the University of Pennsylvania and in Washington, DC.

He has coauthored two books, *The Cydonia Codex* (2005) and *The Martian Codex* (2009), as well as six science papers related to anomalous formations on Mars, which were published in peer-reviewed science journals.

Haas has been a guest speaker at numerous conferences and appeared on the Gaia Network's *Beyond Belief* and the History Channel's *Ancient Aliens, The Proof Is Out There,* and *The UnXplained* with William Shatner.